B

Monographs in Mathematics
Vol. 80

Edited by
A. Borel
J. Moser
S.-T. Yau

Birkhäuser
Boston · Basel · Stuttgart

Enrico Giusti

Minimal Surfaces and Functions of Bounded Variation

1984

Birkhäuser
Boston · Basel · Stuttgart

Birkhäuser extends thanks to the Department of Mathematics, of the Australian National University for permission to publish this volume part of which originally appeared under its sponsorship in 1977.

Library of Congress Cataloging in Publication Data

Giusti, Enrico.
    Minimal surfaces and functions of bounded
variation.

    Bibliography: p.
    Includes index.
    1. Surfaces, Minimal.   2. Functions of bounded
variation.   I. Title.
QA644.G53   1984   516.3'62   84-11195
ISBN 0-8176-3153-4
ISBN 3-7643-3153-4

CIP-Kurztitelaufnahme der Deutschen Bibliothek

Giusti, Enrico:
Minimal surfaces and functions of bounded
variation / Enrico Giusti.
Basel ; Boston ; Stuttgart :
    (Monographs in mathematics ; 80)
    ISBN 3-7643-3153-4

NE: GT

© Birkhäuser Boston, Inc., 1984
Printed in Switzerland
ISBN 0-8176-3153-4
ISBN 3-7643-3153-4

Table of Contents

Introduction

The problem of finding minimal surfaces, i.e. of finding the surface of least area among those bounded by a given curve, was one of the first considered after the foundation of the calculus of variations, and is one which received a satisfactory solution only in recent years. Called the problem of Plateau, after the blind physicist who did beautiful experiments with soap films and bubbles, it has resisted the efforts of many mathematicians for more than a century. It was only in the thirties that a solution was given to the problem of Plateau in 3-dimensional Euclidean space, with the papers of Douglas [DJ] and Radò [RT1, 2]. The methods of Douglas and Radò were developed and extended in 3-dimensions by several authors, but none of the results was shown to hold even for minimal hypersurfaces in higher dimension, let alone surfaces of higher dimension and codimension.

It was not until thirty years later that the problem of Plateau was successfully attacked in its full generality, by several authors using measure-theoretic methods; in particular see De Giorgi [DG1, 2, 4, 5], Reifenberg [RE], Federer and Fleming [FF] and Almgren [AF1, 2]. Federer and Fleming defined a k-dimensional surface in $\mathbb{R}^n$ as a k-current, i.e. a continuous linear functional on k-forms. Their method is treated in full detail in the splendid book of Federer [FH1]. A different attitude was taken by Almgren [AF1, 2] (and Allard [AW]) who introduced k-dimensional varifolds, i.e. Radon measures on $\mathbb{R}^n \times G(n, k)$, where $G(n, k)$ denotes the Grassmann manifold of k-planes in $\mathbb{R}^n$. On the other hand, the ideas of De Giorgi [DG5] were never published in widely circulated journals. They were known to the experts but the work was not really available to a larger audience. In the formalism of De Giorgi, a hypersurface in $\mathbb{R}^n$ is a boundary of a measurable set E whose characteristic function φ_E has distributional derivatives that are Radon measures of (locally) finite total variation (briefly Caccioppoli sets). In this case the $(n-1)$-dimensional area is taken as the total variation of $D\varphi_E$. It is not difficult to show the existence of a solution for Plateau's problem in some weak sense. It is a much more difficult task to prove that the hypersurfaces so obtained (and in general every hypersurface locally minimizing area) are actually regular, except possibly for a closed singular set. This is the central result of the above mentioned paper of De Giorgi [DG5] and it was later simplified and completed by Miranda [MM3] who showed that the singular set Σ has zero $(n-1)$-dimensional measure.

The main idea of [DG5] is the following. For every $x \in \partial E$ it is possible to define an approximate normal vector

$$v_\rho(x) = \int\limits_{B_\rho} D\varphi_E \Big/ \int\limits_{B_\rho} |D\varphi_E|.$$

One can show that if, for some $x \in \partial E$ and some $\rho > 0$, the vector $v_\rho(x)$ has length close enough to 1, then the difference

$$1 - |v_r(x)|$$

converges to 0 as $r \to 0$. This is the hardest part of the proof; it is then quite easy to show that ∂E is regular (analytic) in a neighbourhood of x.

The method of proof outlined above is very powerful when looking for regularity "almost-everywhere". The reader can compare the results of [DG5] with the techniques of [AF1] and Chapter 5 of [FH1], and with the theory of non-linear elliptic systems as developed in [MC2], [GM], [GE1].

Once the regularity almost-everywhere has been established, it is natural to ask whether the singular set Σ can exist at all. One is then naturally led to study the behaviour of ∂E near a point, say the origin, and this is done by blowing up the situation, i.e. by considering the sets

$$E_k = \left\{ x \in \mathbb{R}^n : \frac{x}{k} \in E \right\} \quad k = 1, 2, \ldots$$

Due to the geometric invariance of the area, all the sets E_k are minimal and a subsequence will converge in measure to a set C, which is itself minimal. Moreover, C is a cone, roughly speaking a tangent cone to E at 0. One can see that E is regular near 0 if and only if ∂C is a hyperplane, so that the existence of singularities in ∂E is reduced to the existence of singular minimal cones.

In [AF1], Almgren proved the non-existence of singular minimal cones in $\mathbb{R}^4$, and in [SJ], Simons extended this result up to dimension seven, thus proving the regularity of minimal hypersurfaces in $\mathbb{R}^n$, for $n \le 7$. This result is the best possible since the cone

$$S = \{ x \in \mathbb{R}^8 : x_1^2 + x_2^2 + x_3^2 + x_4^2 < x_5^2 + x_6^2 + x_7^2 + x_8^2 \}$$

is a singular minimal cone in $\mathbb{R}^8$. This was shown by Bombieri, De Giorgi and Giusti [BDGG].

Finally, starting from the result of Simons, Federer [FH2] proved that the dimension of the singular set cannot exceed $n - 8$.

Thus far, we have discussed only general (parametric) minimal hypersurfaces. A special case of primary interest occurs when we ask that our surface E be the graph of a function $u(x)$ defined in some open set Ω. Such surfaces are usually called non-parametric minimal surfaces.

It is customary in the literature on the subject to consider open sets $\Omega \subset \mathbb{R}^n$, so that our surface lies in the cylinder $Q = \Omega \times \mathbb{R} \subset \mathbb{R}^{n+1}$; and has dimension n. Although sometimes it may cause some confusion, in particular when applying parametric methods to the non-parametric case, we prefer to retain this well-established rule.

If $u: \Omega \to \mathbb{R}$ is a smooth function, the area of its graph is given by

$$\mathscr{A}(u) = \int_\Omega \sqrt{1 + |Du|^2}\, dx,$$

and therefore u minimizes the area if and only if it is a solution of the minimal surface equation

$$D_i \left\{ \frac{D_i u}{\sqrt{1 + |Du|^2}} \right\} = 0 \text{ in } \Omega$$

or more briefly

$$\operatorname{div} T(u) = 0; \quad T(u) = Du(1 + |Du|^2)^{-1/2}.$$

A natural question is that of the existence of solutions of the Dirichlet problem; namely of solutions of the minimal surface equation taking prescribed values on the boundary of Ω.

A peculiarity of the minimal surface equation (as compared for instance with Laplace's equation) is that this problem is not generally solvable. When $n = 2$, it was proved by Bernstein [BS1] (and in increasing generality by Haar [HA] and Radò [RT1]) that a solution exists for arbitrary data if Ω is convex, but may fail to exist without the convexity of the domain, even if the boundary datum φ is C^∞ and has arbitrarily small absolute value (Finn [F1], Jenkins and Serrin [JS2]). In the same paper, Jenkins and Serrin proved that the Dirichlet problem in n dimensions is always solvable if the mean curvature of $\partial\Omega$ is nowhere negative. We shall give here a proof of this result, following the ideas of [GD], [SG] and [MM4], without using the theory of non-linear elliptic equations.

The unpleasant restriction on the mean curvature of $\partial\Omega$ can be avoided by a suitable generalization of the Dirichlet problem. More precisely, we do not impose the boundary condition $u = \varphi$ as a characterization of the class of function competing to minimize the area $\mathscr{A}(u)$, but rather we introduce it in the functional under consideration as a penalization, and we look for a minimum of

$$\mathscr{J}(u) = \int_{\Omega} \sqrt{1 + |Du|^2}\, dx + \int_{\partial\Omega} |u - \varphi|\, dH_{n-1}.$$

It is easily seen that a solution of the Dirichlet problem also minimizes $\mathscr{J}$; on the other hand, the new functional always has a minimum in the class $BV(\Omega)$, of functions with bounded variation in Ω, independently of the mean curvature of the boundary.

In general the minimizing function will not take the value φ on $\partial\Omega$; however this will happen at every point $x_0 \in \partial\Omega$ where the mean curvature of $\partial\Omega$ is non negative and φ is continuous.

The connection between parametric and non-parametric minimal surfaces is given by a theorem due to M. Miranda [MM2], which states that u minimizes the area $\mathscr{A}(u)$ in Ω if and only if its subgraph

$$U = \{(x, t) \in Q = \Omega \times \mathbb{R};\ t < u(x)\}$$

is a set of least perimeter in Q. As a consequence, the regularity results for parametric minimal surfaces, in conjunction with the a priori estimates for the gradient of Bombieri, De Giorgi and Miranda [BDGM] (later simplified by Trudinger [TN1] and generalized in [TN2] and [LU2]) imply the regularity everywhere of non-parametric minimal hypersurfaces in arbitrary dimension.

Another important point in the theory of non-parametric minimal surfaces is the so-called Bernstein problem. In 1915, S. Bernstein proved that the affine functions $u(x) = \langle a, x \rangle + b$ are the only entire solutions of the minimal surface equation in the plane [BS2]. Several proofs of the Beinstein's theorem were found later (we shall discuss two of them), but all are restricted to the two-dimensional case. It was only in 1962 that Fleming [FW] gave a different proof, suitable for extension to higher dimensions.

Roughly speaking, Fleming's idea was the following. Let u be an entire solution, and let U be its subgraph. We can consider the sets

$$U_j = \{x \in \mathbb{R}^{n+1} : jx \in U\}$$

and show that a subsequence converges to some minimal set C. Fleming then proved that C is a cone and that its boundary ∂C is a hyperplane if and only if u was an affine function. In other words, the existence of nontrivial entire minimal graphs in $\mathbb{R}^n$ implies the existence of singular minimal cones in $\mathbb{R}^n$; or better in $\mathbb{R}^{n-1}$, as was shown later by De Giorgi [DG6]. The above mentioned results of Almgren and Simons extend the Bernstein theorem up to dimension 7. This result is the best possible since there exist entire solutions of the minimal surface equation in $\mathbb{R}^8$, that are not affine [BDGG].

The aim of these notes is to present a theory of minimal hypersurfaces in arbitrary dimension along the lines sketched above. The book is almost self-contained, the only prerequisites being a fairly good knowledge of general measure theory and some familiarity with the theory of elliptic partial differential equations. For the reader's convenience, the essential results with appropriate references are gathered in Appendix C.

The book consists in two parts, following the distinction between parametric and non-parametric minimal surfaces. The first part repeats with minor changes the notes of a course given at the Australian National University in the spring of 1976, and published there with the assistance of Graham Williams [GE3], who moreover wrote the section 1.30 and the Appendix A, and revised the English throughout.

The first four chapters are devoted to the theory of BV spaces, i.e. spaces of functions whose distributional derivatives are Radon measures of locally bounded total variation. We discuss the general properties of these spaces, emphasizing compactness, semicontinuity, approximation with smooth functions and traces (compare [MM1], [MM5], [AGMMP]). As a particular case we study sets E whose characteristic functions are in the space BV. These are known as Caccioppoli sets or sets of locally finite perimeter. We introduce the notion of reduced boundary ∂^*E as the set of those points of ∂E for which there exists a tangent hyperplane to ∂E, and we discuss its properties, in particular the regularity.

The next four chapters (Chapters 5–8) are dedicated to the proof of the regularity almost-everywhere of minimal hypersurfaces. This is the main part of the notes and contains the theorem of De Giorgi [DG5]. In Chapters 9 and 10 we deal with the problem of minimal cones. In particular, in Chapter 9 we prove the existence of tangent cones to a minimal hypersurface by means of the blow-up procedure described above. Also we prepare the way for the proof of Simons' result [SJ] by reducing it to the problem of the existence of minimal cones with only one singular point, the vertex. In the next chapter it is proved that such cones cannot exist in $\mathbb{R}^n$ for $n \leq 7$ thus proving Simons' theorem. The proof presented here is due to Leon Simon, and does not make use of the differential geometric methods of the original proof of Simons. Instead it involves the differential operators δ introduced by Miranda [MM4] and a careful use of test functions.

The eleventh chapter deals with the dimension of the singular set Σ and contains the proof of the result of Federer mentioned above [FH2] that the dimension of Σ cannot exceed $n - 8$.

The second part is devoted to non-parametric minimal surfaces. The first two chapters (Chapters 12 and 13) are dedicated to the classical Dirichlet problem for the minimal surface equation. In particular we prove the a priori estimate for the gradient, and we discuss the relations between boundary mean curvature and general solvability of the Dirichlet problem.

Then follow three chapters in which the same problem is treated in its relaxed formulation, in the space BV of functions of bounded variation. We consider the Dirichlet problem with L^1 and also with infinite boundary data, and we discuss the existence and uniqueness of the solution, as well as regularity in the interior and at the boundary.

Finally, Chapter 17 is entirely dedicated to the Bernstein problem.

Part 1
Parametric Minimal Surfaces

1. Functions of Bounded Variation and Caccioppoli Sets

In this chapter we introduce the notions of functions of bounded variation and Caccioppoli sets. We derive some important properties which will be of use in later chapters and towards the end of this chapter we obtain an existence theorem for minimal sets.

Frequent use will be made of the following spaces of functions:

$$C^m(\Omega), \ C_0^m(\Omega), \ L^p(\Omega), \ W^{1,1}(\Omega).$$

Their definitions and properties may be found in many text books (for example [AR]). We also make use of k-dimensional Hausdorff measure, H_k, especially when $k = n - 1$ and the reader should be familiar with the definitions of these measures (see page 128. A complete treatment can be found in [FH1] and [RC1]).

1.1 Definition: Let $\Omega \subseteq \mathbb{R}^n$ be an open set and let $f \in L^1(\Omega)$. Define

$$\int_\Omega |Df| = \sup \left\{ \int_\Omega f \operatorname{div} g \, dx : g = (g_1, \ldots, g_n) \in C_0^1(\Omega; \mathbb{R}^n) \right.$$

$$\text{and} \quad |g(x)| \leq 1 \quad \text{for} \quad x \in \Omega \Big\},$$

where $\operatorname{div} g = \sum_{i=1}^{n} \dfrac{\partial g_i}{\partial x_i}$.

1.2 Example: If $f \in C^1(\Omega)$, then integration by parts gives

$$\int_\Omega f \operatorname{div} g \, dx = -\int_\Omega \sum_{i=1}^{n} \frac{\partial f}{\partial x_i} g_i \, dx$$

for every $g \in C_0^1(\Omega; \mathbb{R}^n)$, so that

$$\int_\Omega |Df| = \int_\Omega |\operatorname{grad} f| \, dx,$$

where

$$\operatorname{grad} f = \left(\frac{\partial f}{\partial x_1}, \frac{\partial f}{\partial x_2}, \ldots, \frac{\partial f}{\partial x_n} \right).$$

More generally, if f belongs to the Sobolev space $W^{1,1}(\Omega)$, then

$$\int_{\Omega} |Df| = \int_{\Omega} |\operatorname{grad} f| \, dx$$

where now $\operatorname{grad} f = (f_1, \ldots, f_n)$ and $f_1, \ldots, f_n$ are the generalized derivatives of f.

1.3 Definition: A function of $f \in L^1(\Omega)$ is said to have *bounded variation* in Ω if $\int_{\Omega} |Df| < \infty$. We define $BV(\Omega)$ as the space of all functions in $L^1(\Omega)$ with bounded variation.

1.4 Example: It can be seen from Example 1.2 that $W^{1,1}(\Omega) \subseteq BV(\Omega)$. The fact that the two spaces are not equal can be seen from the next example.

Suppose $E \subseteq \mathbb{R}^n$ has C^2 boundary and consider φ_E, the characteristic function of E, which is defined by

$$\varphi_E(x) = \begin{cases} 1 & \text{if } x \in E \\ 0 & \text{if } x \in \mathbb{R}^n - E. \end{cases}$$

If in addition E is bounded, then

$$\int_{\Omega} \varphi_E dx = |E \cap \Omega| = \text{Lebesgue measure of } E \cap \Omega$$

and $\varphi_E \in L^1(\Omega)$. However, φ_E does not belong to $W^{1,1}(\Omega)$.

Suppose $g \in C_0^1(\Omega; R^n)$. Then, by the Gauss-Green theorem,

$$\int \varphi_E \operatorname{div} g dx = \int_E \operatorname{div} g dx = \int_{\partial E} g \cdot v dH_{n-1},$$

where $v(x)$ is the outward unit normal to ∂E at x and H_{n-1} is $(n-1)$-dimensional Hausdorff measure.

Now $|v(x)| = 1$, so that, if $|g(x)| \leq 1$ and $g \in C_0^1(\Omega; R^n)$, then

$$\int_{\partial E} g \cdot v dH_{n-1} \leq H_{n-1}(\partial E \cap \Omega).$$

and hence

$$\int_{\Omega} |D\varphi_E| = \sup\left\{\int_{\Omega} \varphi_E \operatorname{div} g\,dx : g \in C_0^1(\Omega; \mathbb{R}^n),\ |g(x)| \leq 1\right\} \leq$$

$$\leq H_{n-1}(\Omega \cap \partial E) < \infty.$$

Thus $\varphi_E \in BV(\Omega)$ and in fact

(1.1) $\int_{\Omega} |D\varphi_E| = H_{n-1}(\partial E \cap \Omega).$

To prove (1.1) we need only show that

$$\int_{\Omega} |D\varphi_E| \geq H_{n-1}(\partial E \cap \Omega).$$

Since E has C^2 boundary, $v(x)$ will be a C^1 vector valued function of x with $|v(x)| = 1$ and so may be extended to a function N, defined on the whole of $\mathbb{R}^n$, such that $N \in C^1(\mathbb{R}^n; \mathbb{R}^n)$ and $|N(x)| \leq 1$ for all x. Now if $\eta \in C_0^\infty(\Omega)$ and $|\eta| \leq 1$, then we have, setting $g = N\eta$,

$$\int_E \operatorname{div} g\,dx = \int_{\partial E} \eta\,dH_{n-1}$$

so that

$$\int_{\Omega} |D\varphi_E| \geq \sup\left\{\int_{\partial E} \eta\,dH_{n-1} : \eta \in C_0^\infty(\Omega),\ |\eta| \leq 1\right\} = H_{n-1}(\partial E \cap \Omega).$$

1.5   Remark: If $f \in BV(\Omega)$ and Df is the gradient of f in the sense of distributions (see [MCB1]), then Df is a vector valued Radon measure and $\int_{\Omega} |Df|$ is the total variation of Df on Ω. Thus we may extend the definition of $\int_A |Df|$ to include cases where $A \subset \Omega$ is not necessarily open.

In Example 1.4 we considered a particular class of functions in $BV(\Omega)$, namely the characteristic functions of sets with smooth boundaries. We now extend these ideas to more general sets.

1.6   Definition: Let E be a Borel set and Ω an open set in $\mathbb{R}^n$. Define the *perimeter of E in Ω* as

$$P(E, \Omega) = \int_\Omega |D\varphi_E| = \sup\{\int_E \operatorname{div} g dx : g \in C_0^1(\Omega; \mathbb{R}^n), \ |g(x)| \leq 1\}.$$

If $\Omega = \mathbb{R}^n$, denote $P(E) = P(E, \mathbb{R}^n)$.

If a Borel set E has locally finite perimeter, that is, if $P(E, \Omega) < \infty$ for every bounded open set Ω, then E is called a Caccioppoli set.

1.7 Remark: The following simple properties of Caccioppoli sets may be proved:

(i)   if $\Omega \subseteq \Omega_1$, $P(E, \Omega) \leq P(E, \Omega_1)$

with equality holding when $E \subset \subset \Omega$ (i.e. $\bar{E}$ is a compact subset of Ω)

(ii) $P(E_1 \cup E_2, \Omega) \leq P(E_1, \Omega) + P(E_2, \Omega)$

with equality holding when $\operatorname{dist}(E_1, E_2) > 0$

(iii)   if $|E| = 0$, then $P(E) = 0$ and so, in particular,
         if $|E_1 \Delta E_2| = |(E_1 - E_2) \cup (E_2 - E_1)| = 0$, then $P(E_1) = P(E_2)$.

1.8 Remark: As in Remark 1.5 we see that if E is a Caccioppoli set, then there exists a vector valued Radon measure ω with locally finite variation such that, for all $g \in C_0^1(\Omega; \mathbb{R}^n)$,

(1.2) $\int_E \operatorname{div} g dx = \int g \cdot d\omega$

where $\omega = -D\varphi_E$.

In fact the converse is true. Suppose there exists a vector valued Radon measure ω such that (1.2) holds. Then, if $g \in C_0^1(\Omega; \mathbb{R}^n)$ and $|g(x)| \leq 1$,

$$\int_E \operatorname{div} g dx = \int g \cdot d\omega \leq |\omega|(\Omega) < \infty,$$

where $|\omega|(\Omega)$ is the total variation of ω on Ω. Thus $P(E, \Omega) \leq |\omega|(\Omega) < \infty$ for each bounded open set Ω and E is a Caccioppoli set with $\omega = -D\varphi_E$.

Note also that

(1.3) $\operatorname{spt} D\varphi_E \subseteq \partial E$,

where

$$spt\, D\varphi_E = \mathbb{R}^n - \cup \{\text{open sets } A \text{ such that } g \in C_0^1(A;\ \mathbb{R}^n) \Rightarrow \int g \cdot D\varphi_E = 0\}.$$

Indeed, if $x \notin \partial E$, then there must exist an open set A such that $x \in A$ and either $A \subseteq E$ or $A \subseteq \mathbb{R}^n - E$. If $A \subseteq \mathbb{R}^n - E$, then $\varphi_E = 0$ on A; hence if $g \in C_0^1(A;\ \mathbb{R}^n)$, then $\int g \cdot D\varphi_E = -\int \varphi_E \,\mathrm{div}\, g\, dx = 0$. If $A \subseteq E$, then $\varphi_E = 1$ on A; hence if $g \in C_0^1(A;\ \mathbb{R}^n)$, then $\int g \cdot D\varphi_E = -\int \varphi_E \,\mathrm{div}\, g\, dx = \int \mathrm{div}\, g\, dx = 0$. Thus $x \in A \subseteq \mathbb{R}^n - spt\, D\varphi_E$ and (1.3) holds.

Now using (1.3) we may write (1.2) as

$$(1.4) \quad \int_E \mathrm{div}\, g\, dx = -\int_{\partial E} g \cdot D\varphi_E;$$

that is, a Gauss-Green formula holds in a generalized sense for each Caccioppoli set E, and in fact, by the remarks above, Caccioppoli sets are characterized by this property.

One of the most important properties of BV functions is demonstrated by the next theorem.

1.9   Theorem (*Semicontinuity*): *Let* $\Omega \subseteq \mathbb{R}^n$ *be an open set and* $\{f_j\}$ *a sequence of functions in* $BV(\Omega)$ *which converge in* $L^1_{loc}(\Omega)$ *to a function f. Then*

$$(1.5) \quad \int_\Omega |Df| \leq \liminf_{j \to \infty} \int_\Omega |Df_j|.$$

Proof: Let $g \in C_0^1(\Omega;\ \mathbb{R}^n)$ be such that $|g| \leq 1$. Then

$$\int f \,\mathrm{div}\, g\, dx = \lim_{j \to \infty} \int f_j \,\mathrm{div}\, g\, dx \leq \liminf_{j \to \infty} \int_\Omega |Df_j|.$$

Now (1.5) follows on taking the supremum over all such g.   □

1.10   Example: From equation (1.1) of Example 1.4 we see that if E is a set in $\mathbb{R}^n$ with smooth (C^2) boundary, then $P(E, \Omega) = H_{n-1}(\partial E \cap \Omega)$. However, if E is not smooth this may not be true as the following example shows.

Let $\{x_i\}$ be the sequence of all rational points in $\mathbb{R}^n$ and let

$$B_i = B(x_i, 2^{-i}) = \{x \in \mathbb{R}^n : |x - x_i| < 2^{-i}\}$$

and $E = \bigcup_{i=0}^{\infty} B_i$. Then

$$|E| \leq \sum_{i=0}^{\infty} |B_i| = \omega_n \sum_{i=0}^{\infty} 2^{-in} = \frac{\omega_n}{1 - 2^{-n}}$$

where $\omega_n = \Gamma\left(\frac{1}{2}\right)^n / \Gamma\left(\frac{n}{2} + 1\right)$ is the measure of the unit ball in $\mathbb{R}^n$. However, the rational points are dense in $\mathbb{R}^n$ and so $\bar{E} = \mathbb{R}^n$. Thus $|\partial E| = \infty$ which implies that $H_{n-1}(\partial E) = \infty$.

On the other hand, if we define $E_k = \bigcup_{i=1}^{k} B_i$, then $E_k \to E$ (or $\varphi_{E_k} \to \varphi_E$ in $L^1(\mathbb{R}^n)$) and, as ∂E_k is piecewise smooth, we may apply (1.1) to obtain

$$P(E_k) = H_{n-1}(\partial E_k) \leq H_{n-1}\left(\bigcup_{i=0}^{k} \partial B_i\right) = n\omega_{n-1} \sum_{i=0}^{k} 2^{-i(n-1)}$$

$$\leq \frac{n\omega_{n-1}}{1 - 2^{-(n-1)}}.$$

Now from Theorem 1.9

$$P(E) \leq \liminf_{k \to \infty} P(E_k) \leq \frac{n\omega_{n-1}}{1 - 2^{-(n-1)}} < \infty.$$

1.11   Example: This example shows that equality need not be achieved in (1.5). Let $\Omega = (0, 2\pi) \subseteq \mathbb{R}^1$ and $f_j(x) = \frac{1}{j} \sin jx$ for $x \in \Omega$ and $j = 1, 2, \ldots$. The f_j are in $L^1(\Omega)$ and furthermore

$$\int_{\Omega} |f_j| \, dx = \frac{1}{j} \int_{0}^{2\pi} |\sin jx| \, dx \leq \frac{2\pi}{j} \to 0.$$

Thus $f_j \to 0$ in $L^1(\Omega)$. On the other hand, as the f_j are smooth,

$$\int_{\Omega} |Df_j| = \int_{0}^{2\pi} |\cos jx| \, dx = 4.$$

Although, generally, we cannot expect equality, we can prove it in certain special cases. For example see Propositions 1.13 and 1.15.

1.12 Remark: Under the norm

$$\|f\|_{BV} = \|f\|_{L^1} + \int_{\Omega} |Df|,$$

$BV(\Omega)$ is a Banach space.

The norm properties follow easily from the definitions of $\|f\|_{L^1}$ and $\int_{\Omega} |Df|$ and so it only remains to prove completeness. Suppose $\{f_j\}$ is a Cauchy sequence in $BV(\Omega)$; then, by the definition of the norm, it must also be a Cauchy sequence in $L^1(\Omega)$ and hence, by the completeness of $L^1(\Omega)$, there exists a function f in $L^1(\Omega)$ such that $f_j \to f$ in $L^1(\Omega)$. Since $\{f_j\}$ is a Cauchy sequence in $BV(\Omega)$, $\|f_i\|_{BV}$ is bounded. Thus $\int_{\Omega} |Df_i|$ is bounded as $j \to \infty$ and so, by the semi-continuity Theorem 1.9, $f \in BV(\Omega)$. It remains only to show that $f_j \to f$ in $BV(\Omega)$ or, since we already have convergence in $L^1(\Omega)$, that

$$\int_{\Omega} |D(f_j - f)| \to 0 \text{ as } j \to \infty.$$

Suppose $\varepsilon > 0$; then there exists an integer N such that

$$j, k \geq N \Rightarrow \|f_j - f_k\|_{BV} < \varepsilon$$

$$\Rightarrow \int_{\Omega} |D(f_j - f_k)| < \varepsilon$$

Now $f_k \to f$ in $L^1(\Omega)$ and so $f_j - f_k \to f_j - f$ in $L^1(\Omega)$. Thus by Theorem 1.9

$$\int_{\Omega} |D(f_j - f)| \leq \lim_{k \to \infty} \inf \int_{\Omega} |D(f_j - f_k)| \leq \varepsilon.$$

Since $\varepsilon > 0$ was arbitrary, $f_j \to f$ in $BV(\Omega)$.

1.13 Proposition: *Suppose $\{f_j\}$ is a sequence in $BV(\Omega)$ such that $f_j \to f$ in $L^1_{loc}(\Omega)$ and*

$$\lim_{j \to \infty} \int_{\Omega} |Df_j| = \int_{\Omega} |Df|.$$

Then for every open set $A \subseteq \Omega$

(1.6) $\int\limits_{\bar{A} \cap \Omega} |Df| \geq \limsup\limits_{j \to \infty} \int\limits_{\bar{A} \cap \Omega} |Df_j|.$

In particular, if $\int\limits_{\partial A \cap \Omega} |Df| = 0$, then

$$\int\limits_{A} |Df| = \lim\limits_{j \to \infty} \int\limits_{A} |Df_j|.$$

Proof: Let $B = \Omega - \bar{A}$ so that B is open. Then, by Theorem 1.9,

$$\int\limits_{A} |Df| \leq \liminf\limits_{j \to \infty} \int\limits_{A} |Df_j|$$

$$\int\limits_{B} |Df| \leq \liminf\limits_{j \to \infty} \int\limits_{B} |Df_j|.$$

On the other hand,

$$\int\limits_{\bar{A} \cap \Omega} |Df| + \int\limits_{B} |Df| = \int\limits_{\Omega} |Df|$$

$$= \lim\limits_{j \to \infty} \int\limits_{\Omega} |Df_j|$$

$$\geq \limsup\limits_{j \to \infty} \int\limits_{\bar{A} \cap \Omega} |Df_j| + \liminf\limits_{j \to \infty} \int\limits_{B} |Df_j|$$

$$\geq \limsup\limits_{j \to \infty} \int\limits_{\bar{A} \cap \Omega} |Df_j| + \int\limits_{B} |Df|$$

and (1.6) follows. $\square$

1.14 Symmetric Mollifiers: A function $\eta(x)$ is called a *mollifier* if

(i) $\eta(x) \in C_0^\infty(\mathbb{R}^n)$,
(ii) η is zero outside a compact subset of $B_1 = \{x \in \mathbb{R}^n : |x| < 1\}$,
(iii) $\int \eta(x) dx = 1$.

If in addition we have

     (iv) $\eta(x) \geq 0$
     (v) $\eta(x) = \mu(|x|)$ for some function $\mu : \mathbb{R}^+ \to \mathbb{R}$,

then η is a *positive symmetric mollifier*.
    For example the function

$$\gamma(x) = \begin{cases} 0 & |x| \geq 1 \\ C\exp\left(\dfrac{1}{|x|^2 - 1}\right) & |x| < 1, \end{cases}$$

where C is a constant chosen so that $\int \gamma(x)dx = 1$, is a positive symmetric mollifier.
    Given such an η and a function $f \in L^1_{loc}(\mathbb{R}^n)$, define for each $\varepsilon > 0$

$$\eta_\varepsilon(x) = \varepsilon^{-n}\eta\left(\frac{x}{\varepsilon}\right)$$

$$f_\varepsilon = \eta_\varepsilon * f;$$

that is,

$$(1.7) \quad f_\varepsilon(x) = \varepsilon^{-n}\int_{\mathbb{R}^n} \eta\left(\frac{x-z}{\varepsilon}\right)f(z)dz = (-1)^n\varepsilon^{-n}\int_{\mathbb{R}^n}\eta\left(\frac{y}{\varepsilon}\right)f(x-y)dy$$

$$= \int_{\mathbb{R}^n}\eta(w)f(x+\varepsilon w)dw$$

Then, using the standard properties of mollifiers, we may show

(1.8)   (a) $f_\varepsilon \in C^\infty(\mathbb{R}^n)$, $f_\varepsilon \to f$ in $L^1_{loc}(\mathbb{R}^n)$ and if $f \in L^1(\mathbb{R}^n)$, then $f_\varepsilon \to f$ in $L^1(\mathbb{R}^n)$,
        (b) $A \leq f(x) \leq B$ for all $x \Rightarrow A \leq f_\varepsilon(x) \leq B$ for all x,
        (c) if $f, g \in L^1(\mathbb{R}^n)$, then $\int_{\mathbb{R}^n} f_\varepsilon g \, dx = \int_{\mathbb{R}^n} fg_\varepsilon \, dx$,
        (d) if $f \in C^1(\mathbb{R}^n)$, then $\dfrac{\partial f_\varepsilon}{\partial x_i} = \left(\dfrac{\partial f}{\partial x_i}\right)_\varepsilon$,
        (e) $\operatorname{spt} f \subseteq A \Rightarrow \operatorname{spt} f_\varepsilon \subseteq A_\varepsilon = \{x : \operatorname{dist}(x, A) \leq \varepsilon\}$,

where the support of f (denoted $\operatorname{spt} f$) is defined by

$$\operatorname{spt} f = \operatorname{closure}\{x \in \mathbb{R}^n : f(x) \neq 0\}.$$

1.15   **Proposition:** *Suppose $f \in BV(\Omega)$ and suppose $A \subset \subset \Omega$ is an open set such that*

(1.9) $\int\limits_{\partial A} |Df| = 0.$

Then, if f_ε are the mollified functions described above (where f is extended to be 0 outside Ω if necessary),

(1.10) $\int\limits_{A} |Df| = \lim\limits_{\varepsilon \to 0} \int\limits_{A} |Df_\varepsilon| \, dx.$

Proof: Since $f_\varepsilon \to f$ in $L^1(\Omega)$, we already have, by Theorem 1.9, the inequality

$$\int\limits_{A} |Df| \leq \liminf\limits_{\varepsilon \to 0} \int\limits_{A} |Df_\varepsilon| \, dx$$

and so it remains only to prove a reverse inequality.

    Suppose $g \in C_0^1(A; \mathbb{R}^n)$ and $|g| \leq 1$; then, by the properties described in 1.14,

$$\int\limits_{\Omega} f_\varepsilon \operatorname{div} g \, dx = \int\limits_{\Omega} f (\operatorname{div} g)_\varepsilon \, dx$$

$$= \int\limits_{\Omega_1} f \operatorname{div} g_\varepsilon \, dx.$$

Now

$$|g| \leq 1 \Rightarrow |g_\varepsilon| \leq 1$$

and

$$\operatorname{spt} g \subseteq A \Rightarrow \operatorname{spt} g_\varepsilon \subseteq A_\varepsilon = \{x : \operatorname{dist}(x, A) \leq \varepsilon\}$$

so that

$$\int\limits_{\Omega} f_\varepsilon \operatorname{div} g \, dx \leq \int\limits_{A_\varepsilon} |Df|.$$

On taking the supremum over all such g, we see that

$$\int\limits_{A} |Df_\varepsilon| \, dx \leq \int\limits_{A_\varepsilon} |Df|.$$

Thus

$$\limsup_{\varepsilon \to 0} \int_A |Df_\varepsilon|\, dx \le \lim_{\varepsilon \to 0} \int_{A_\varepsilon} |Df| = \int_A |Df|$$

and so by (1.9)

$$\limsup_{\varepsilon \to 0} \int_A |Df_\varepsilon|\, dx \le \int_A |Df|. \qquad\qquad \square$$

1.16   Remark: If $A = \mathbb{R}^n$, then 1.15 shows that

$$\int_{\mathbb{R}^n} |Df| = \lim_{\varepsilon \to 0} \int_{\mathbb{R}^n} |Df_\varepsilon|\, dx$$

and, in particular, if $f = \varphi_E$, a characteristic function, then we obtain

$$P(E) = \lim_{\varepsilon \to 0} \int_{\mathbb{R}^n} |D(\varphi_E)_\varepsilon|\, dx.$$

This equality could be used as the definition of $P(E)$ and indeed a definition only slightly different was used by De Giorgi ([DG1]). His definition replaced f_ε by

$$f_t(x) = \int g_t(y - x) f(y)\, dy$$

for $f \in L^1(\mathbb{R}^n)$, where

$$g_t(y) = (\pi t)^{-n/2} e^{-|y|^2/t}.$$

Thus η_ε has been replaced by g_t which although not actually a mollifier does possess many of the properties described in 1.14. It is possible to show that, if $f \in L^1(\mathbb{R}^n)$, then the function

$$t \to \int_\Omega |Df_t|\, dx \quad (t > 0)$$

is a decreasing function of t and we can define

$$\int |Df| = \lim_{t \to 0^+} \int |Df_t|\, dx$$

and

$$P(E) = \int |D\varphi_E| = \lim_{t \to 0^+} \int |D(\varphi_E)_t| \, dx.$$

It is easy to prove that this definition coincides with our original one.

We are now in a position to show that every function f in $BV(\Omega)$ can be approximated, in some sense, by C^∞ functions. Approximation in the BV-norm cannot be expected since the closure of the C^∞ functions in this norm is the Sobolev space $W^{1,1}(\Omega)$, which we have shown in 1.4 not to be equal to $BV(\Omega)$. So, in particular, we cannot expect to find $f_j \in C^\infty(\Omega)$ such that $f_j \to f$ in $L^1(\Omega)$ and $\int_\Omega |D(f_j - f)| \to 0$.

1.17   Theorem [AG]: *Let $f \in BV(\Omega)$. Then there exists a sequence $\{f_j\}$ in $C^\infty(\Omega)$ such that*

$$\lim_{j \to \infty} \int_\Omega |f_j - f| \, dx = 0$$

(1.11)

$$\lim_{j \to \infty} \int_\Omega |Df_j| \, dx = \int_\Omega |Df|.$$

Proof: Let $\varepsilon > 0$. There exists a number m such that if we set

$$\Omega_k = \left\{ x \in \Omega : \mathrm{dist}(x, \partial\Omega) > \frac{1}{m+k} \right\}; \quad k = 0, 1, 2, \ldots$$

then

$$\int_{\Omega - \Omega_0} |Df| < \varepsilon.$$

Consider now the sets A_i, $i = 1, 2, \ldots$, defined by $A_1 = \Omega_2$ and for $i = 2, 3, \ldots$

$$A_i = \Omega_{i+1} - \bar{\Omega}_{i-1}$$

and let $\{\varphi_i\}$ be a partition of the unity subordinate to the covering $\{A_i\}$; that is

$$\varphi_i \in C_0^\infty(A_i), \quad 0 \leq \varphi_i \leq 1, \quad \sum_{i=1}^\infty \varphi_i = 1.$$

Let η be a positive symmetric mollifier as defined in 1.14. For every index i we can choose $\varepsilon_i > 0$ such that

(1.12) $\operatorname{spt} \eta_{\varepsilon_i} * (f\varphi_i) \subset \Omega_{i+2} - \bar\Omega_{i-2} \quad (\Omega_{-1} = \varnothing)$

(1.13) $\int |\eta_{\varepsilon_i} * (f\varphi_i) - f\varphi_i| dx < \varepsilon 2^{-i}$

(1.14) $\int |\eta_{\varepsilon_i} * (fD\varphi_i) - fD\varphi_i| dx < \varepsilon 2^{-i}.$

Finally, let

$$f_\varepsilon = \sum_{i=1}^\infty \eta_{\varepsilon_i} * (f\varphi_i).$$

It follows from (1.12) that the sum defining f_ε is locally finite, and hence $f_\varepsilon \in C^\infty(\Omega)$. Moreover, since $f = \sum_{i=1}^\infty f\varphi_i$, we have from (1.13)

$$\int_\Omega |f_\varepsilon - f| dx \leq \sum_{i=1}^\infty \int_\Omega |\eta_{\varepsilon_i} * (f\varphi_i) - f\varphi_i| dx < \varepsilon,$$

and therefore, when $\varepsilon \to 0$, f_ε converges to f in $L^1(\Omega)$. From Theorem 1.9 we have

(1.15) $\int_\Omega |Df| \leq \liminf_{\varepsilon \to 0} \int_\Omega |Df_\varepsilon|.$

Let now $g \in C_0^1(\Omega; \mathbb{R}^n)$, with $|g| \leq 1$. We have:

$$\int f_\varepsilon \operatorname{div} g \, dx = \sum_{i=1}^\infty \int \eta_{\varepsilon_i} * (f\varphi_i) \operatorname{div} g \, dx = \sum_{i=1}^\infty \int f\varphi_i \operatorname{div}(\eta_{\varepsilon_i} * g) dx$$

and hence:

$$\int f_\varepsilon \operatorname{div} g \, dx = \int f \operatorname{div}(\varphi_1 \eta_{\varepsilon_1} * g) dx + \sum_{i=2}^\infty \int f \operatorname{div}(\varphi_i \eta_{\varepsilon_i} * g) dx -$$

$$- \sum_{i=1}^\infty \int \langle g, \eta_{\varepsilon_i} * (fD\varphi_i) - fD\varphi_i \rangle dx$$

where we have used the identity $\sum\limits_{i=1}^{\infty} D\varphi_i = 0$.

Since $|\varphi_i\eta_{\varepsilon_i} * g| \leq 1$ we get:

$$\int \mathrm{div}(\varphi_1\eta_{\varepsilon_i} * g)dx \leq \int\limits_{\Omega} |Df|,$$

whereas taking into account the fact that the intersection of more than any three of the sets A_i is empty:

$$\sum\limits_{i=2}^{\infty} \int f \mathrm{div}(\varphi_i\eta_{\varepsilon_i} * g)dx \leq 3 \int\limits_{\Omega-\Omega_0} |Df| < 3\varepsilon.$$

From (1.14) we obtain therefore the inequality

$$\int f_\varepsilon \mathrm{div}\, g\, dx \leq \int\limits_{\Omega} |Df| + 4\varepsilon,$$

and recalling the Definition 1.1.:

$$\int\limits_{\Omega} |Df_\varepsilon| \leq \int\limits_{\Omega} |Df| + 4\varepsilon.$$

Letting $\varepsilon \to 0$ and comparing with (1.15) we obtain at once the conclusion of the theorem.   □

1.18   Remark: For every $\varepsilon > 0$, for every $N > 0$ and for every $x_0 \in \partial\Omega$,

(1.15) $\lim\limits_{\rho \to 0} \rho^{-N} \int\limits_{B(x_0,\rho)\cap\Omega} |f_\varepsilon - f|\, dx = 0,$

where $B(x_0, \rho) = \{x \in \mathbb{R}^n : |x - x_0| < \rho\}$.

In fact, if $x \in B(x_0, \rho) \cap \Omega$ then, by our choice of spt φ_k,

$$f_\varepsilon(x) - f(x) = \sum\limits_{k=k_0}^{\infty} [\eta_{\varepsilon_k} * (f\varphi_k) - f\varphi_k],$$

where $k_0 = \left[\dfrac{1}{\rho}\right] - m - 2$. Then from (1.13)

$$\int_{B(x_0, \rho) \cap \Omega} |f_\varepsilon - f| \, dx \leqq 2^{-k_0} C$$

and, using the relationship between k_0 and ρ, (1.15) follows.

We can now prove another important theorem for $BV(\Omega)$, namely a compactness result.

1.19   Theorem (*Compactness*): *Let Ω be a bounded open set in $\mathbb{R}^n$ which is sufficiently regular for the Rellich Theorem to hold (see [AR], [MCB1]; of course it is sufficient that the boundary of Ω is Lipschitz-continuous). Then sets of functions uniformly bounded in BV-norm are relatively compact in $L^1(\Omega)$.*

Proof: Suppose $\{f_j\}$ is a sequence in $BV(\Omega)$ such that $\|f_j\|_{BV} \leqq M$. For each j, by Theorem 1.17, we can choose $\tilde{f}_j \in C^\infty(\Omega)$ such that

$$\int_\Omega |\tilde{f}_j - f_j| < \frac{1}{j} \quad \text{and} \quad \|\tilde{f}_j\|_{BV} \leqq M + 2.$$

Now, by the Rellich Theorem ([AR], [MCB1]), $\{\tilde{f}_j\}$ is relatively compact in $L^1(\Omega)$, so there is a subsequence converging in $L^1(\Omega)$ to a function f. By Theorem 1.9, $f \in BV(\Omega)$, and is the limit of a sequence extracted from $\{f_j\}$.   □

Notice that with exactly the same argument we could prove that bounded sets in $BV(\Omega)$ are relatively compact in $L^p(\Omega)$ for any p such that $1 \leqq p < \frac{n}{n-1}$.

Using this last theorem, together with the semicontinuity Theorem 1.9, it is now an easy matter to prove the existence of minimizing Caccioppoli sets.

1.20   Theorem (*Existence of minimal surfaces [DG5]*): *Let Ω be a bounded open set in $\mathbb{R}^n$ and let L be a Caccioppoli set. Then there exists a set E coinciding with L outside Ω and such that*

$$\int |D\varphi_E| \leqq \int |D\varphi_F|$$

for every set F with $F = L$ outside Ω.

Proof: Since Ω is bounded, there exists a number R such that $\Omega \subset\subset B_R = \{x \in \mathbb{R}^n : |x| < R\}$. Now, if $F = L$ outside Ω, then

$$\int |D\varphi_F| = \int_{B_R} |D\varphi_F| + \int_{\mathbb{R}^n - B_R} |D\varphi_L|.$$

So we need only show that there exists a set E in B_R coinciding with L outside Ω such that

$$\int_{B_R} |D\varphi_E| \leq \int_{B_R} |D\varphi_F|$$

for each set F in B_R coinciding with L outside Ω.

Obviously $\int\limits_{B_R} |D\varphi_F|$ is bounded below by 0 and so, if $\{E_j\}$ is a minimizing sequence, we must have that $\int\limits_{B_R} |D\varphi_{E_j}|$ is uniformly bounded. Furthermore B_R is bounded so $\int\limits_{B_R} |\varphi_{E_j}| dx$ is uniformly bounded. Hence φ_{E_j} is a bounded sequence in $BV(B_R)$ and, by Theorem 1.19, there must exist a subsequence, still denoted $\{\varphi_{E_j}\}$, which converges in $L^1(B_R)$ to a function f. Since $\varphi_{E_j}(x) \to f(x)$ for almost all x in B_R and $\varphi_{E_j}(x)$ is either 0 or 1 we may assume that f is the characteristic function of a set E which coincides with L outside Ω. Now, by the semicontinuity Theorem 1.9, we see that E must provide the required minimum.   □

1.21 Remark: (i) In some sense the set L determines boundary values for E. Roughly speaking, ∂E minimizes the area among all surfaces with boundary $\partial L \cap \partial \Omega$.

For example, in $\mathbb{R}^2$ let

$$\Omega = B_2 = \{x \in \mathbb{R}^2 : |x| < 2\},$$

$$L = \{x \in \mathbb{R}^2 : (x_1)^2 + (x_2 - 1)^2 < 4\}.$$

Then E will be the set

$$\left\{ (x_1, x_2) \in L : x_2 > \frac{1}{2} \right\}$$

(ii) From the proof of the theorem it is obvious that the nature of L far from E (in fact, in the proof, the nature of L outside B_R) has no effect on the set $E \cap \Omega$ which we are interested in.

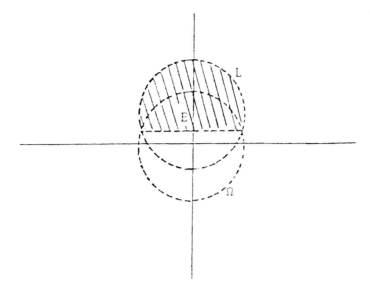

1.22   Remark: (i) The set Ω can act as an obstacle forcing ∂E away from the minimal surface spanning $\partial L \cap \partial \Omega$.

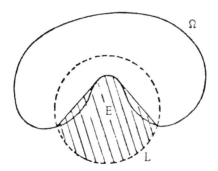

(ii) By the same methods we could also minimize functionals of the form

$$\int |D\varphi_E| + \int_E H(x)\,dx.$$

Having proved this existence theorem, the remainder of our work will be concerned with the smoothness of the resulting minimal set and with various other properties of Cacciopoli sets and BV functions.

The next theorem, although concerning BV functions, will be especially useful in obtaining suitable smooth approximations for Cacciopoli sets.

1.23 **Theorem** (*Coarea formula*) [*FR*], [*DG4*]: *Let* $f \in BV(\Omega)$ *and define*

$$F_t = \{x \in \Omega : f(x) < t\}.$$

Then

(1.16) $\quad \int\limits_{\Omega} |Df| = \int\limits_{-\infty}^{\infty} dt \int\limits_{\Omega} |D\varphi_{F_t}|.$

Proof: Suppose $g \in C_0^1(\Omega; \mathbb{R}^n)$ and $|g| \leq 1$. We first consider the case where $f \geq 0$. Noting that

$$f(x) = \int\limits_0^{\infty} (1 - \varphi_{F_t}(x)) dt,$$

we have

$$\int f \operatorname{div} g \, dx = \int dx \int\limits_0^{\infty} (1 - \varphi_{F_t}(x)) \operatorname{div} g \, dt$$

$$= \int\limits_0^{\infty} dt \{ \int \operatorname{div} g \, dx - \int \varphi_{F_t} \operatorname{div} g \, dx \}.$$

Now $g \in C_0^1(\Omega; \mathbb{R}^n)$ so that

$$\int \operatorname{div} g \, dx = 0.$$

Thus

$$\int f \operatorname{div} g \, dx = -\int\limits_0^{\infty} dt \int\limits_{F_t} \operatorname{div} g \, dx \leq \int\limits_0^{\infty} dt \int\limits_{\Omega} |D\varphi_{F_t}|.$$

If $f \leq 0$ we note that

$$f(x) = \int\limits_{-\infty}^{0} \varphi_{F_t}(x) dt$$

and obtain, as above,

$$\int f \operatorname{div} g \, dx \leq \int\limits_{-\infty}^{0} dt \int\limits_{\Omega} |D\varphi_{F_t}|.$$

Splitting f into positive and negative part gives

$$\int f \operatorname{div} g \, dx \leqq \int\limits_{-\infty}^{\infty} dt \int\limits_{\Omega} |D\varphi_{F_t}|$$

for any function $f \in BV(\Omega)$. On taking the supremum over all g such that $g \in C_0^1(\Omega; \mathbb{R}^n)$ and $|g| \leq 1$, we have

$$\int\limits_{\Omega} |Df| \leq \int\limits_{-\infty}^{\infty} dt \int\limits_{\Omega} |D\varphi_{F_t}|.$$

To prove the reverse inequality, assume first that (1.16) holds if $f \in C^\infty(\Omega)$. Now take $f \in BV(\Omega)$ and let $\{f_j\}$ be the approximating sequence from Theorem 1.17. Then, since $f_j \to f$ in $L^1(\Omega)$ and

$$\int\limits_{\Omega} |f_j - f| \, dx = \int\limits_{-\infty}^{\infty} dt \int |\varphi_{F_{jt}} - \varphi_{F_t}| \, dx,$$

where $F_{jt} = \{x \in \Omega : f_j(x) < t\}$, there must exist a subsequence, still denoted $\{f_j\}$, such that

$$\varphi_{F_{jt}} \to \varphi_{F_t} \quad \text{in } L^1(\Omega)$$

for almost all t. Hence, by Fatou's lemma and the semicontinuity Theorem 1.9,

$$\int\limits_{\Omega} |Df| = \lim_{j \to \infty} \int\limits_{\Omega} |Df_j| = \lim_{j \to \infty} \int\limits_{-\infty}^{\infty} dt \int\limits_{\Omega} |D\varphi_{F_{jt}}| \geq \int\limits_{-\infty}^{\infty} dt \int\limits_{\Omega} |D\varphi_{F_t}|.$$

Thus it remains only to prove (1.16) for $f \in C^\infty(\Omega)$. We can repeat, exactly, the above reasoning but now approximate $f \in C^\infty(\Omega)$ by continuous piecewise linear functions f_j. Now we show (1.16) holds for continuous piecewise linear functions. Suppose $\Omega = \bigcup\limits_{i=1}^{N} \Omega_i$ and

$$f(x) = \langle c_i, x \rangle + b_i \quad \text{if} \quad x \in \Omega_i,$$

where $c_i \in \mathbb{R}^n$ and $b_i \in \mathbb{R}$. Then f is in $W^{1,1}(\Omega)$ and

$$\int\limits_{\Omega} |Df| = \sum_{i=1}^{N} |c_i| |\Omega_i|.$$

Furthermore,

$$\int_{\Omega_i} |D\varphi_{F_t}| = H_{n-1}(\{x \in \Omega_i : f(x) = t\})$$

$$= H_{n-1}(\{x \in \Omega_i : \langle c_i, x \rangle + b_i = t\}).$$

Now assume the x_1-axis is perpendicular to the hyperplane $\{x \in \Omega_i : \langle c_i, x \rangle + b_i = t\}$. Thus, on introducing a change of variables to the right hand side of (1.16), we have

$$\int_{-\infty}^{\infty} dt \int_{\Omega_i} |D\varphi_{F_t}| = \int_{-\infty}^{\infty} |c_i| H_{n-1}(\{x \in \Omega_i : x_1 = t\}) dt = |c_i| |\Omega_i|.$$

Hence

$$\int_{-\infty}^{\infty} dt \int_{\Omega} |D\varphi_{F_t}| = \sum_{i=1}^{N} |c_i| |\Omega_i| = \int_{\Omega} |Df|. \qquad \square$$

We can now prove a theorem analogous to 1.17, but in this case we approximate Caccioppoli sets rather than BV functions. The proof uses a lemma which will be proved after the theorem for ease of presentation.

1.24 Theorem: *Every bounded Caccioppoli set E can be approximated by a sequence of C^∞ sets E_j, such that*

$$\int |\varphi_{E_j} - \varphi_E| dx \to 0, \quad \int |D\varphi_{E_j}| \to \int |D\varphi_E|.$$

Proof: By Theorem 1.17 we know that φ_E may be approximated by a sequence of C^∞ functions obtained by mollifying φ_E. From these functions we obtain the required approximating sets by using 1.23.

Let $\varepsilon > 0$ and let $f_\varepsilon = \eta_\varepsilon * \varphi_E$ as in 1.14; then by 1.23, noting that $0 \le f_\varepsilon \le 1$,

$$\int |Df_\varepsilon| = \int_0^1 dt \int |D\varphi_{E_{\varepsilon t}}|$$

where $E_{\varepsilon t} = \{x : f_\varepsilon(x) < t\}$. By Remark 1.16

$$\int |D\varphi_E| = \lim_{\varepsilon \to 0} \int |Df_\varepsilon| dx.$$

Suppose $\varepsilon_j \to 0$ as $j \to \infty$; then, by Lemma 1.25 (to follow), for every t with $0 < t < 1$

$$\varphi_{E_{jt}} \to \varphi_E \quad \text{almost everywhere}$$

$(E_j = E_{\varepsilon_j})$. In fact the convergence holds in $L^1(\mathbb{R}^n)$. Then by Theorem 1.9

$$\liminf_{j \to \infty} \int |D\varphi_{E_{jt}}| \geq \int |D\varphi_E| \quad \text{for each } t.$$

Thus

$$\int |D\varphi_E| = \lim_{j \to \infty} \int |Df_{\varepsilon_j}| \geq \int_0^1 dt \liminf_{j \to \infty} \int |D\varphi_{E_{jt}}| \geq \int |D\varphi_E|,$$

and so, for almost all t in $(0, 1)$,

$$\liminf_{j \to \infty} \int |D\varphi_{E_{jt}}| = \int |D\varphi_E|.$$

Moreover, by Sard's lemma [MB], for almost every t, ∂E_{jt} is regular and so we may choose a $t \in (0, 1)$ such that if $F_j = E_{jt}$ then

(i) ∂F_j is smooth
(ii) $\varphi_{F_j} \to \varphi_E$ in $L^1(\mathbb{R}^n)$,
(iii) $\int |D\varphi_E| = \liminf_{j \to \infty} \int |D\varphi_{F_j}|$.

Taking a subsequence of $\{\varepsilon_j\}$, we can ensure that (iii) holds with $\lim_{j \to \infty}$ rather than $\liminf_{j \to \infty}$ and the theorem is proved. □

1.25   Lemma: *Let $0 < t < 1$, $\varepsilon_j \to 0$ as $j \to \infty$ and*

$$E_j = \{x \in \mathbb{R}^n : f_{\varepsilon_j}(x) > t\},$$

where $f_{\varepsilon_j} = \eta_{\varepsilon_j} * \varphi_E$. Then

$$\lim_{j \to \infty} \int |\varphi_{E_j} - \varphi_E| dx \leq \frac{1}{\min(t, 1-t)} \int |f_{\varepsilon_j} - \varphi_E| dx.$$

Proof: By definition

$$f_{\varepsilon_j} - \varphi_E > t \quad \text{in } E_j - E$$

$$\varphi_E - f_{\varepsilon_j} \geqq 1 - t \quad \text{in } E - E_j,$$

so that

$$\int |f_{\varepsilon_j} - \varphi_E| dx \geqq \int_{E_j - E} |f_{\varepsilon_j} - \varphi_E| dx + \int_{E - E_j} |f_{\varepsilon_j} - \varphi_E| dx \geqq$$

$$\geqq t|E_j - E| + (1 - t)|E - E_j| \geqq$$

$$\geqq \min(t, 1 - t) \int |\varphi_{E_j} - \varphi_E| dx. \qquad \square$$

1.26   Remark: By Proposition 1.13, if $\int_{\partial A} |D\varphi_E| = 0$ where A is an open set, then

$$\int_A |D\varphi_E| = \lim_{j \to \infty} \int_A |D\varphi_{E_j}|.$$

1.27   Remark: It is not, in general, possible to approximate a Caccioppoli set E by C^∞ sets contained inside E (nor is it possible from the outside). For example, consider the set E of 1.10. If F is any set containing E, we must have $\bar{F} = \mathbb{R}^n$ and hence either $|F - E| = \infty$ or $|\partial F| = \infty$ (or both). It follows then that E cannot be approximated from the outside by smooth sets.

The following theorem is an extension of the Sobolev inequality for functions in $W^{1,1}(\Omega)$.

1.28   Theorem: *(Sobolev inequality) (A). Let $f \in BV(\mathbb{R}^n)$ have compact support. Then*

$$(1.17) \quad \left(\int |f|^{\frac{n}{n-1}} dx\right)^{\frac{n-1}{n}} \leqq c_1 \int |Df|,$$

where c_1 is a constant depending only on n.
   (B) Let $f \in BV(B_\rho)$ and define

$$f_\rho = \frac{1}{|B_\rho|} \int_{B_\rho} f dx.$$

Then

$$(1.18) \quad \left(\int_{\dot B_\rho} |f - f_\rho|^{\frac{n}{n-1}} dx \right)^{\frac{n-1}{n}} \leqq c_2 \int_{\dot B_\rho} |Df|,$$

where c_2 is a constant depending only on n.

Proof: Inequalities (1.17) and (1.18) are well known for $f \in C_0^\infty(\mathbb{R}^n)$, $C^\infty(B_\rho)$ respectively (for example see [AR]). Consider (1.17) and choose a sequence $\{f_j\}$ in $C_0^\infty(\mathbb{R}^n)$ such that $f_j \to f$ in $L^1(\mathbb{R}^n)$ and $\int |Df_j| \to \int |Df|$. Now, by (1.17) for $C_0^\infty(\mathbb{R}^n)$, the functions f_j are uniformly bounded in the $L^{\frac{n}{n-1}}(\mathbb{R}^n)$ norm, and so a subsequence will converge weakly to some function $f_0 \in L^{\frac{n}{n-1}}(\mathbb{R}^n)$. However, $f_j \to f$ in $L^1(\mathbb{R}^n)$; accordingly $f_0 = f$ and $f_j \to f$ weakly in $L^{\frac{n}{n-1}}(\mathbb{R}^n)$ so that

$$\left(\int |f|^{\frac{n}{n-1}} dx \right)^{\frac{n-1}{n}} \leqq \liminf_{j \to \infty} \left(\int |f_j|^{\frac{n}{n-1}} dx \right)^{\frac{n-1}{n}} \leqq c_1 \lim_{j \to \infty} \int |Df_j| = c_1 \int |Df|.$$

Similarly we may prove (1.18). $\qquad\qquad\qquad\qquad\qquad\qquad$ □

1.29 Corollary (*Isoperimetric inequalities*): *Let E be a bounded Caccioppoli set in $\mathbb{R}^n$. Then*

$$(1.19) \quad |E|^{\frac{n-1}{n}} \leqq c_1(n) \int |D\varphi_E|$$

$$(1.20) \quad \min\{|E \cap B_\rho|, |(\mathbb{R}^n - E) \cap B_\rho|\}^{\frac{n-1}{n}} \leqq c_2(n) \int_{B_\rho} |D\varphi_E|.$$

Proof: Inequality (1.19) follows immediately from (1.17) by noting that $\varphi_E \in BV(\mathbb{R}^n)$ and has compact support.

Let $f = \varphi_E$. Then $f_\rho = \dfrac{|E \cap B_\rho|}{|B_\rho|}$ and so

$$\int_{\dot B_\rho} |f - f_\rho|^{\frac{n}{n-1}} dx = \left\{ 1 - \frac{|E \cap B_\rho|}{|B_\rho|} \right\}^{\frac{n-1}{n}} |E \cap B_\rho| +$$

$$+ \left\{ \frac{|E \cap B_\rho|}{|B_\rho|} \right\}^{\frac{n-1}{n}} |(\mathbb{R}^n - E) \cap B_\rho|.$$

Hence, noting that

$$|(\mathbb{R}^n - E) \cap B_\rho| = |B_\rho| \left\{ 1 - \frac{|E \cap B_\rho|}{|B_\rho|} \right\},$$

we have

$$\left\{ \int_{B_\rho} |f - f_\rho|^{\frac{n}{n-1}} dx \right\}^{\frac{n-1}{n}} \geq \min \left\{ |E \cap B_\rho|, |(\mathbb{R}^n - E) \cap B_\rho| \right\}^{\frac{n-1}{n}}$$

$$\frac{1}{|B_\rho|} \left\{ |E \cap B_\rho|^{\frac{n}{n-1}} + |(\mathbb{R}^n - E) \cap B_\rho|^{\frac{n}{n-1}} \right\}^{\frac{n-1}{n}} \geqq$$

$$\geqq \min \left\{ |E \cap B_\rho|, |(\mathbb{R}^n - E) \cap B_\rho| \right\}^{\frac{n-1}{n}}$$

and (1.20) follows.

1.30   Other Definitions of BV functions: There are several other definitions of the "space of functions of bounded variation" especially in the case of functions from $\mathbb{R}$ to $\mathbb{R}$. We now give a few of the most common definitions and discuss their relationship with the one we have used.

Suppose $f: \mathbb{R} \to \mathbb{R}$ and $a < b$ are real numbers. Then define the *variation* of f on $[a, b]$ as

$$V_a^b(f) = \sup \left\{ \sum_{i=1}^{m} |f(t_i) - f(t_{i-1})| : m \in \mathbb{N} \text{ and } a = t_0 < t_1 < \ldots < t_m = b \right\}$$

The function f is said to be of *bounded variation* if there exists a constant K such that

$$V_a^b(f) \leqq K \quad \text{for all } a < b$$

and we define the *variation* of f as

$$V(f) = \sup \{ V_a^b(f) : a < b \text{ real numbers}).$$

This definition is well known and frequently used for functions $f: \mathbb{R} \to \mathbb{R}$. However, in the situations we are considering it does suffer some drawbacks. We are mostly interested in integrals and measures and so we want changes on sets of measure zero to have no effect. Unfortunately if we change the value of a function, even at just one point, the variation will be affected. To overcome these problems we assume that our functions are integrable and define a quantity known as the essential variation.

The essential variation depends on a notion called approximate continuity. We say; f is *approximately continuous* at x_0, if for all $\varepsilon > 0$, the set $A = \mathbb{R} - f^{-1}\{y: |y - f(x_0)| > \varepsilon\}$ has density 0 at x_0 where the *density* of a set A at a point x_0 is defined as

$$\lim_{r \to 0} \frac{|A \cap \{x: |x - x_0| < r\}|}{|\{x: |x - x_0| < r\}|} = \lim_{t \to 0} \frac{|A \cap B(x_0, t)|}{|B(x_0, t)|}.$$

In the usual definition of continuity we would require that dist$\{x_0, A) > 0$. Thus the definition of approximate continuity relaxes this to allow some points of A close to x_0 but not very many. An important property of this definition is that, provided we leave the value of f at x_0 unchanged, the approximate continuity will be unaffected by changes of f on a set of measure zero.

Suppose f is integrable and $a < b$ are real numbers. Then we define the *essential variation* of f on (a, b) as

$$\text{ess } V_a^b(f) = \sup\left\{ \sum_{i=1}^{m} |f(t_i) - f(t_{i-1})| : m \in \mathbb{N} \text{ and } a < t_0 < t_1 \ldots < t_m < b \right.$$

$$\left. \text{where each } t_i \text{ is a point of approximate continuity} \right\}.$$

We say f has *bounded essential variation* or $f \in BV(\mathbb{R})$ if there exists a constant K such that

$$\text{ess } V_a^b(f) \leqq K \quad \text{for all } a < b$$

and define the *essential variation* of f as

$$\text{ess } V(f) = \sup\{\text{ess } V_a^b(f) : a < b \text{ real numbers}\}.$$

It can be shown that this definition of $BV(\mathbb{R})$ is equivalent to ours, that is:

(a) f has bounded essential variation if and only if

$$\sup\{\int_{\mathbb{R}} fg' dx : g \in C_0^1(\mathbb{R}), |g| \leqq 1\} < \infty.$$

Indeed there are other equivalent definitions:

(b) the derivative of f (in the distributional sense) is a finite measure,

(c) f can be approximated in L^1 by C^∞ functions with uniformly bounded variation,

(d) there exists a function g such that g has bounded variation and $f = g$ almost everywhere.

Some of these conditions carry over to $\mathbb{R}^n$ with the appropriate changes made because of the higher dimension and once again they may be shown to be equivalent to our original definition.

(b′) The derivative Df of f (in the distributional sense) is a finite vector valued measure, (equivalence follows from Remarks 1.5 and 1.8),

(c′) f can be approximated in the L^1 norm by C^∞ functions with uniformly bounded variation, that is

$$f_j \to f \text{ in } L^1(\mathbb{R}^n) \text{ and } \int |Df_j|\,dx \leq K,$$

(equivalence follows from Theorems 1.9 and 1.17).

Although (d) does not carry over directly, it has been shown by Federer (4.5.9 of [FH2] or [FH1]) that, from f we can obtain a function g which has various nice properties and corresponds in the 1-dimensional case to the function of (d) above.

Finally, the definition of bounded essential variation has been extended to $\mathbb{R}^n$ as follows (see 4.5.10 [FH2]):

(a′) $f \in L^1(\mathbb{R}^n)$ has bounded essential variation if, for $i = 1, 2, \ldots, n$, the essential variation of $f(y_1, \ldots, y_{i-1}, t, y_i, \ldots, y_{n-1})$ with respect to t is integrable with respect to y in $(n-1)$-space.

At first glance these definitions may appear to be markedly different but their equivalence is easily established once the following equalities are proved:

$$\sup\left\{ \int_{\mathbb{R}^n} f \operatorname{div} g\,dx : g \in C_0^1(\mathbb{R}^n; \mathbb{R}^n), \ |g| \leq 1 \right\}$$

(b″) $= \int |Df|$

(c″) $= \inf \left\{ \limsup_{j \to \infty} \int_{\mathbb{R}^n} |Df_j|\,dx : \{f_j\} \subseteq C_0^\infty(\mathbb{R}^n), f_j \to f \text{ in } L^1(\mathbb{R}^n) \right\}.$

For the case $n = 1$, we have

(a″) $\int |Df| = \operatorname{ess} V(f)$

(d″) $= \inf\{V(g): g = f \text{ almost everywhere}\}$

and if $n > 1$ we have

(a‴) $\int |Df| = \int\limits_{\mathbb{R}^{n-1}} \text{ess } V(f_{i,y}) dy$ for each $i = 1, 2, \ldots, n,$

where $f_{i,y}(t) = f(y_1, y_2, \ldots, y_{i-1}, t, y_i, \ldots, y_{n-1})$.

2. Traces of *BV* Functions

Since we are regarding $BV(\Omega)$ as a subset of $L^1(\Omega)$ we consider equivalence classes of functions rather than the functions themselves. Thus it makes no sense to talk of the value of a *BV* function on a set of measure zero since we may change the values of the function on such a set without changing its equivalence class. However it is important to be able to talk about the value of a *BV* function on the boundary of a set even though such a boundary may have measure zero. Obviously such a notion of values on a set of measure zero must take into account the value of the function on surrounding sets rather than just the set itself. It is the aim of this chapter to give a rigorous and meaningful definition of the trace of a *BV* function on the boundary of the set and then to develop some of the properties of the trace.

An essential tool in the definition of trace is the following theorem of Lebesgue.

2.1   Theorem (*Lebesgue's Theorem*): *If $f \in L^1(\mathbb{R}^n)$, then for almost all $x \in \mathbb{R}^n$*

$$(2.1) \quad \lim_{\rho \to 0} \rho^{-n} \int_{B_\rho} |f(x+t) - f(x)| \, dt = 0.$$

(For a proof see 2.9.9 of [FH2].)

As $\partial \Omega$ has zero measure, the preceding theorem does not allow us to define unambiguously the values of f on $\partial \Omega$ and in fact a general f in $L^1(\Omega)$ has no trace on $\partial \Omega$. It is precisely the existence of derivatives that makes it possible to define a trace on $\partial \Omega$, and more generally on closed hypersurfaces $S \subset \bar{\Omega}$, for functions $f \in BV(\Omega)$.

The following covering lemma will be useful in this and later chapters.

2.2   Lemma: *Let $A \subseteq \mathbb{R}^n$ and $\rho : A \to (0, 1)$; then there exists a countable set of points $\{x_i\}$ in A such that:*

$$(2.2) \quad B(x_i, \rho(x_i)) \cap B(x_j, \rho(x_j)) = \varnothing \quad \text{if } i \neq j$$

$$(2.3) \quad A \subseteq \bigcup_{i=1}^{\infty} B(x_i, 3\rho(x_i)),$$

where $B(x, r) = \{y \in \mathbb{R}^n : |x - y| < r\}$.

Proof: For $k = 1, 2, \ldots$ let

$$A_k = \{x \in A : 2^{-k} \leqq \rho(x) < 2^{1-k}\}.$$

Consider the class of all subsets L of A_1, such that $B(x, \rho(x)) \cap B(y, \rho(y)) = \emptyset$ for all $x, y \in L, x \neq y$. If we order such sets by inclusion, then by Zorn's lemma there is a maximal subset L_1. If M is any compact subset of $\mathbb{R}^n$, then $L_1 \cap M$ must be finite and so L_1 itself must be countable.

Similarly, let L_2 be a maximal subset of A_2, such that $B(x, \rho(x)) \cap B(y, \rho(y)) = \emptyset$ for all $x, y \in L_1 \cup L_2, x \neq y$. Now proceeding by induction let $L_j \subseteq A_j$ be a maximal subset satisfying:

$$B(x, \rho(x)) \cap B(y, \rho(y)) = \emptyset \text{ for all } x, y \in L_1 \cup L_2 \cup \ldots \cup L_j, \quad x \neq y.$$

The set $A = \bigcup\limits_{i=1}^{\infty} L_i$ is then countable and satisfies

$$B(x, \rho(x)) \cap B(y, \rho(y)) = \emptyset \text{ for all } x, y \in A, \quad x \neq y.$$

We now show that A satisfies (2.3).

Suppose $z \in A$, then, as $A = \bigcup\limits_{k=1}^{\infty} A_k$, there exists a k such that $z \in A_k$ and further there exists an $x \in L_1 \cup L_2 \cup \ldots \cup L_k$ such that $B(x, \rho(x)) \cap B(z, \rho(z)) \neq \emptyset$ (otherwise L_k would not be maximal). On the other hand, by the definition of A_k and L_i,

$$\rho(x) \geqq \frac{1}{2}\rho(z)$$

and hence $z \in B(x, 3\rho(x))$, establishing (2.3). $\square$

In proving future results it will be useful to adopt the following notation:

$$B(x, r) = \{z \in \mathbb{R}^n : |x - z| < r\} = \text{a ball in } \mathbb{R}^n$$

$$\mathscr{B}(y, \rho) = \{t \in \mathbb{R}^{n-1} : |y - t| < \rho\} = \text{a ball in } \mathbb{R}^{n-1}.$$

2.3   Lemma: *Let* $\mathbb{R}_+^n = \{x \in \mathbb{R}^n : x_n > 0\}$ *and let* μ *be a positive Radon measure on* $\mathbb{R}_+^n$ *with* $\mu(\mathbb{R}_+^n) < \infty$. *For* $\rho > 0$ *and* $y \in \mathbb{R}^{n-1} = \partial\mathbb{R}_+^n$, *let*

$$C_\rho^+(y) = \{x \in \mathbb{R}^n : x = (z, t), |y - z| < \rho, 0 < t < \rho\} = \mathscr{B}(y, \rho) \times (0, \rho).$$

Then, for H_{n-1}-almost all $y \in \mathbb{R}^{n-1}$,

(2.4) $$\lim_{\rho \to 0^+} \rho^{1-n} \mu(C_\rho^+(y)) = 0.$$

Proof: Let

$$A_k = \left\{ y \in \mathbb{R}^{n-1} : \limsup_{\rho \to 0} \rho^{1-n} \mu(C_\rho^+(y)) > \frac{1}{k} \right\}.$$

It is sufficient to show that $H_{n-1}(A_k) = 0$ for each k. Given $y \in A_k$ and $\varepsilon > 0$, there exists a number $\rho_y < \varepsilon$ such that

$$\mu(C_{\rho_y}^+(y)) > \frac{1}{2k} \rho_y^{n-1}.$$

Now by the covering Lemma 2.2 we can choose a sequence $y_i \in A_k$ such that the balls $\mathscr{B}(y_i, \rho_i)$, where $\rho_i = \rho_{y_i}$, are disjoint and $A_k \subseteq \bigcup_{i=1}^{\infty} \mathscr{B}(y_i, 3\rho_i)$. Then

$$H_{n-1}(A_k) \leqq \omega_{n-1} \sum_{i=1}^{\infty} (3\rho_i)^{n-1} \leqq 2k\omega_{n-1} 3^{n-1} \sum_{i=1}^{\infty} \mu(C_{\rho_i}^+(y_i)).$$

On the other hand, since $\rho_i < \varepsilon$, we have

$$C_{\rho_i}^+(y_i) \subseteq L_\varepsilon = \{ x \in \mathbb{R}^n : 0 < x_n < \varepsilon \}$$

and hence

$$H_{n-1}(A_k) < 2k\omega_{n-1} 3^{n-1} \mu(L_\varepsilon) \quad \text{for all } \varepsilon > 0.$$

But since $\mu(\mathbb{R}_+^n) < \infty$, we have $\mu(L_\varepsilon) \to 0$ as $\varepsilon \to 0$ and so $H_{n-1}(A_k) = 0$.   □

2.4   Lemma [$MM5$]: *Let*

$$C_R^+ = \mathscr{B}(0, R) \times (0, R) = \mathscr{B}_R \times (0, R)$$

and $f \in BV(C_R^+)$. Then there exists a function $f^+ \in L^1(\mathscr{B}_R)$ such that for H_{n-1}-almost all $y \in \mathscr{B}_R$

(2.5) $\lim\limits_{\rho \to 0} \rho^{-n} \int\limits_{C_\rho^+(y)} |f(z) - f^+(y)| dz = 0.$

Moreover, if $C_R = \mathscr{B}_R \times (-R, R)$, then for every $g \in C_0^1(C_R; \mathbb{R}^n)$

(2.6) $\int\limits_{C_R^+} f \operatorname{div} g \, dx = - \int\limits_{C_R^+} <g, Df> + \int\limits_{\mathscr{B}_R} f^+ g_n dH_{n-1}.$

Proof: Let us suppose first that f is a smooth function, and for $\varepsilon > 0$ let $f_\varepsilon : \mathscr{B}_R \to \mathbb{R}$ be defined by:

(2.7) $f_\varepsilon(y) = f(y, \varepsilon).$

Let $Q_{\varepsilon,\varepsilon'} = \mathscr{B}_R \times (\varepsilon', \varepsilon)$, $0 < \varepsilon' < \varepsilon < R$. We have:

(2.8) $\int\limits_{\mathscr{B}_R} |f_\varepsilon - f_{\varepsilon'}| dH_{n-1} \leqq \int\limits_{Q_{\varepsilon,\varepsilon'}} |D_n f| dx.$

The sequence f_ε is thus a Cauchy sequence in $L^1(\mathscr{B}_R)$, and therefore it converges to a function $f^+ \in L^1(\mathscr{B}_R)$.

Suppose now that $g \in C_0^1(C_R; \mathbb{R}^n)$. Letting $\varepsilon \to 0$ in the identity

$$\int\limits_{Q_{R,\varepsilon}} f \operatorname{div} g \, dx = - \int\limits_{Q_{R,\varepsilon}} <Df, g> dx + \int\limits_{\mathscr{B}_R} f_\varepsilon g_{n\varepsilon} dH_{n-1}$$

we obtain at once (2.6).

To prove (2.5) we observe that

$$\int\limits_{C_\rho^+(y)} |f(z) - f^+(y)| dz = \int\limits_{\mathscr{B}_\rho(y)} d\eta \int\limits_0^\rho |f(\eta, t) - f^+(y)| dt \leqq$$

$$\leqq \int\limits_{\mathscr{B}_\rho(y)} d\eta \int\limits_0^\rho |f(\eta, t) - f^+(\eta)| dt +$$

$$+ \rho \int\limits_{\mathscr{B}_\rho(y)} |f^+(\eta) - f^+(y)| d\eta.$$

But by the Lebesgue Theorem 2.1

(2.9) $\lim\limits_{\rho \to 0} \rho^{1-n} \int\limits_{\mathscr{B}_\rho(y)} |f^+(\eta) - f^+(y)| d\eta = 0$

for H_{n-1}-almost all $y \in \mathscr{B}_R$. On the other hand, by (2.8),

$$\int_0^\rho dt \int_{\mathscr{B}_\rho(y)} |f(\eta, t) - f^+(\eta)| d\eta \leq \rho \int_{C_\rho^+(y)} |Df|.$$

Thus

$$\rho^{-n} \int_{C_\rho^+(y)} |f(z) - f^+(y)| dz \leq \rho^{1-n} \int_{C_\rho^+(y)} |Df| + \rho^{1-n} \int_{\mathscr{B}_\rho(y)} |f^+(\eta) - f^+(y)| d\eta.$$

But, by Lemma 2.3, for H_{n-1}-almost all $y \in \mathscr{B}_R$

$$\rho^{1-n} \int_{C_\rho^+(y)} |Df| \to 0 \quad \text{as } \rho \to 0$$

and so using (2.9) we obtain (2.5).

The theorem is thus proved for $f \in C^\infty(C_R^+) \cap BV(C_R^+)$. Now let $f \in BV(C_R^+)$, and let f_j be the sequence obtained in Theorem 1.17.

From (2.5) applied to f_j and Remark 1.18 we easily obtain:

$$\lim_{\rho \to 0} \rho^{-n} \int_{C_\rho^+(y)} |f(z) - f_j^+(y)| dz = 0$$

for H_{n-1}-almost every $y \in \mathscr{B}_R$ and for every j. In particular all the traces f_j^+ coincide so that defining $f^+ = f_j^+$ we have (2.5). Finally, writing (2.6) for f_j and passing to the limit as $j \to \infty$ we obtain the full result.   □

2.5   Remark: The function f^+ is called the trace of f on $\mathscr{B}_R$ and obviously

$$f^+(y) = \lim_{\rho \to 0} \frac{1}{|C_\rho^+(y)|} \int_{C_\rho^+(y)} f(z) dz.$$

2.6   Proposition: *Let $f \in BV(C_R^+)$ and let $\{f_j\} \subseteq BV(C_R^+)$ be a sequence converging to f in $L^1(C_R^+)$ such that*

$$\lim_{j \to \infty} \int_{C_R^+} |Df_j| = \int_{C_R^+} |Df|.$$

Then

$$\lim_{j \to \infty} f_j^+ = f^+ \quad \text{in } L^1(\mathscr{B}_R).$$

Proof: From (2.8) if $0 < \beta < R$ and $Q_\beta = Q_{R,\beta}$,

$$\frac{1}{\beta} \int_0^\beta dt \int_{\mathscr{B}_R} |f^+(y) - f(y, t)| dy \leq \int_{Q_\beta} |Df|$$

and hence, if

$$f_\beta(y) = \frac{1}{\beta} \int_0^\beta f(y, t) dt,$$

(2.10) $\int_{\mathscr{B}_R} |f^+(y) - f_\beta(y)| dy \leq \int_{Q_\beta} |Df|.$

Now from (2.10)

(2.11) $\int_{\mathscr{B}_R} |f^+(y) - f_j^+(y)| dy \leq \int_{Q_\beta} |Df| + \int_{Q_\beta} |Df_j| + \int_{\mathscr{B}_R} |f_\beta(y) - f_{j,\beta}(y)| dy$

and

$$\int_{\mathscr{B}_R} |f_\beta(y) - f_{j,\beta}(y)| dy \leq \frac{1}{\beta} \int_0^\beta dt \int_{\mathscr{B}_R} |f(y, t) - f_j(y, t)| dy \leq$$

$$\leq \frac{1}{\beta} \int_{C_R^+} |f - f_j| dx \to 0 \text{ as } j \to \infty.$$

Also by Proposition 1.13

$$\lim_{j \to \infty} \int_{Q_\beta} |Df_j| = \int_{Q_\beta} |Df| \text{ for almost every } \beta.$$

Then in conclusion

$$\limsup_{j \to \infty} \int_{\mathscr{B}_R} |f^+ - f_j^+| dy \leq 2 \int_{Q_\beta} |Df|.$$

for almost every β. Letting $\beta \to 0$, we obtain the result.   □

2.7   Remark: If $C_R^- = \mathscr{B}_R \times (-R, 0)$ and $f \in BV(C_R^-)$, then we can define a trace $f^- \in L^1(\mathscr{B}_R)$ which will satisfy theorems analogous to 2.5 and 2.6.

2.8   Proposition: *Let $f_1 \in BV(C_R^+)$ and $f_2 \in BV(C_R^-)$. Define a function $f: C_R \to \mathbb{R}$ by*

$$f = \begin{cases} f_1 & in \ C_R^+ \\ f_2 & in \ C_R^-. \end{cases}$$

Then $f \in BV(C_R)$ and

$$(2.12) \quad \int_{\mathscr{B}_R} |f^+ - f^-| dH_{n-1} = \int_{\mathscr{B}_R} |Df|.$$

Proof: From (2.6) and the analogue for f_2.

$$\int_{C_R} f \operatorname{div} g \, dx = - \int_{C_{\mathring{R}}} \langle g, Df \rangle - \int_{C_{\bar{\rho}}} \langle g, Df \rangle + \int_{\mathscr{B}_R} (f^+ - f^-) g_n dH_{n-1}.$$

As the right hand side is bounded if $|g| \leq 1$, the function f is in $BV(C_R)$ and

$$\int_{C_R} f \operatorname{div} g \, dx = - \int_{C_R} <g, Df> = - \int_{C_{\mathring{R}}} \langle g, Df \rangle - \int_{C_{\bar{R}}} \langle g, Df \rangle - \int_{\mathscr{B}_R} \langle g, Df \rangle$$

so that

$$- \int_{\mathscr{B}_R} \langle g, Df \rangle = \int_{\mathscr{B}_R} (f^+ - f^-) g_n dH_{n-1} \qquad \text{for } g \in C_0^1(C_R; \mathbb{R}^n)$$

and (2.12) follows. $\square$

2.9   **Remark:** This theorem illustrates an important extension property for *BV* functions and (2.12) shows that unless $f^+ = f^-$ we could not expect a similar theorem for functions in $W^{1,1}$.

We now turn to the case of a general Ω in $\mathbb{R}^n$. Suppose Ω is an open set in $\mathbb{R}^n$, with Lipschitz continuous boundary $\partial \Omega$, and $f \in BV(\Omega)$. If $x_0 \in \partial \Omega$, then with a translation and a rotation we can reduce our considerations to the case $x_0 = 0$ and further suppose that in a neighbourhood of $x_0 = 0$ we have

$$\partial \Omega = \{(y, t) \in \mathbb{R}^n : y \in A, \ t = \omega(y)\}$$

where A is a neighbourhood of 0 in $\mathbb{R}^{n-1}$ and $\omega : A \to \mathbb{R}$ is a Lipschitz continuous function such that if $(y, t) \in \Omega$ then $t > \omega(y)$. If we set

$$\xi = (\eta, \tau) = (y, t - \omega(y))$$

and

$$g(\xi) = f(\eta, \tau + \omega(\eta)) = f(y, t),$$

then the function g is in $BV(C_R^+)$, for some $R > 0$.

If g^+ is the trace of g on $\mathcal{B}_R$, then we define the *trace of f* in $S_R = \{\xi = (\eta, \omega(\eta)): \eta \in \mathcal{B}_R\}$ as

$$\varphi(\xi) = f^+(\eta, \omega(\eta)) = g^+(\eta).$$

Thus, possibly using a partition of unity we have

2.10 **Theorem:** *Let Ω be a bounded open set in $\mathbb{R}^n$ with Lipschitz continuous boundary $\partial\Omega$ and let $f \in BV(\Omega)$. Then there exists a function $\varphi \in L^1(\partial\Omega)$ such that for H_{n-1}-almost all $x \in \partial\Omega$*

$$(2.13) \quad \lim_{\rho \to 0} \rho^{-n} \int_{B_\rho(x) \cap \Omega} |f(z) - \varphi(x)| \, dz = 0.$$

Moreover, for every $g \in C_0^1(\mathbb{R}^n; \mathbb{R}^n)$,

$$(2.14) \quad \int_\Omega f \operatorname{div} g \, dx = -\int_\Omega \langle g, Df \rangle + \int_{\partial\Omega} \varphi \langle g, \vec{n} \rangle \, dH_{n-1}$$

where $\vec{n}$ is the unit outer normal to $\partial\Omega$.
    Similarly from 2.6 we may obtain

2.11 **Theorem:** *Let Ω be a bounded open set in $\mathbb{R}^n$ with Lipschitz continuous boundary $\partial\Omega$, and let f_j, f be functions in $BV(\Omega)$ satisfying*

$$\lim_{j \to \infty} \int_\Omega |f_j - f| \, dx = 0$$

$$\lim_{j \to \infty} \int_\Omega |Df_j| = \int_\Omega |Df|.$$

Then, if φ_j and φ are the traces of f_j and f respectively, we have

$$\lim_{j \to \infty} \int_{\partial\Omega} |\varphi_j - \varphi| \, dH_{n-1} = 0.$$

2.12 Remark: From the definition of trace, Theorem 1.17 and Remark 1.18 it follows that, if $\partial\Omega$ is Lipschitz continuous, then for every $f \in BV(\Omega)$ there exists a sequence $\{f_j\}$ in $C^\infty(\Omega)$ such that

$$\lim_{j\to\infty} \int_\Omega |f_j - f|\,dx = 0$$

$$\lim_{j\to\infty} \int_\Omega |Df_j|\,dx = \int_\Omega |Df|$$

and moreover the trace of each f_j on $\partial\Omega$ coincides with the trace of f.

2.13 Remark: If $A \subset\subset \Omega$ is an open set with Lipschitz continuous boundary ∂A, then $f|_A$ and $f|_{\Omega - \bar{A}}$ (belonging to $BV(A)$ and $BV(\Omega - \bar{A})$ respectively) will have traces on ∂A which we call f_A^- and f_A^+ respectively. Then

$$\lim_{\rho\to0} \rho^{-n} \int_{B_\rho(x)\cap A} |f(z) - f_A^-(x)|\,dz = 0 \text{ for } H_{n-1}\text{- almost all } x \in \partial A,$$

$$\lim_{\rho\to0} \rho^{-n} \int_{B_\rho(x) - A} |f(z) - f_A^+(x)|\,dz = 0 \text{ for } H_{n-1}\text{-almost all } x \in \partial A,$$

and, as in Proposition 2.8,

$$(2.15) \quad \int_{\partial A} |f_A^+ - f_A^-|\,dH_{n-1} = \int_{\partial A} |Df|.$$

Moreover, from the proof of Proposition 2.8, we see that

$$Df = (f_A^+ - f_A^-)v\,dH_{n-1} \text{ on } \partial A,$$

where v is the unit outer normal.

A special case occurs when $\Omega = B_R$ and $A = B_\rho$ with $\rho < R$. The traces are denoted f_ρ^+ and f_ρ^- and so from (2.15) it follows that for almost every ρ

$$f_\rho^+(x) = f_\rho^-(x) = f(x) \quad H_{n-1}\text{-almost everywhere in } \partial B_\rho.$$

Also

$$f_\rho^-(\rho x) = \lim_{\substack{t\to\rho^- \\ t\notin N}} f(tx) \quad \text{in } L^1(\partial B_1),$$

where N is a set of measure zero, and similarly for $f_\rho^+(\rho x)$.

In particular given $x \in B_R$, let $\rho = |x|$ to obtain $f^+(x) = f_\rho^+(x)$. If we do this for each x in B_R we obtain a function f^+ (or f^-) equal almost everywhere to $f(x)$ but now satisfying

$$\lim_{\rho \to 0} \rho^{-n} \int_{B_\rho(x) - B_{|x|}} |f^+(z) - f^+(x)| \, dz = 0$$

for all x (and similarly for $f^-(x)$).

However we do not use this property in what follows.

2.14 Remark: If A and Ω are open sets with $A \subseteq \Omega$, ∂A Lipschitz continuous and $f \in BV(A)$, then we can define a function $F : \Omega \to \mathbb{R}$ by

$$F(x) = \begin{cases} f(x) & x \in A \\ 0 & x \in \Omega - A. \end{cases}$$

Then $F_A^- = f_A^-$ and $F_A^+ = 0$, so that from (2.15)

$$\int_\Omega |DF| = \int_A |Df| + \int_{\partial A \cap \Omega} |f_A^-| \, dH_{n-1}.$$

In particular, if $A = B_\rho$, $\Omega = B_R (\rho < R)$, E is a Caccioppoli set and $f = \varphi_E$, we have, for almost every ρ (and more particularly those ρ for which $\varphi_{E,\rho}^- = \varphi_E$ H_{n-1}-almost everywhere on ∂B_ρ)

(2.16) $P(E \cap B_\rho, B_R) = P(E, B_\rho) + H_{n-1}(\partial B_\rho \cap E)$.

Similarly putting $A = B_R - \bar{B}_\rho$, $\Omega = B_R$

(2.17) $P(E - \bar{B}_\rho, B_R) = P(E, B_R - \bar{B}_\rho) + H_{n-1}(\partial B_\rho \cap E)$

and

(2.18) $P(E \cup \bar{B}_\rho, B_R) = P(B_R - (E \cup \bar{B}_\rho), B_R) =$

$$= P((B_R - E) \cap (B_R - \bar{B}_\rho), B_R) =$$

$$= P(E, B_R - \bar{B}_\rho) + H_{n-1}(\partial B_\rho - E).$$

for almost all $\rho < R$.

We conclude this section by proving a converse of Theorem 2.10.

2.15   Proposition $[GA]$: *Let φ be a function in $L^1(\mathscr{B}_R)$ with compact support. For every $\varepsilon > 0$ there exists a function $f \in W^{1,1}(C_R^+)$ with trace φ on $\mathscr{B}_R$ and such that:*

$$(2.19) \quad \int_{C_R^+} |f| \, dz \leqq \varepsilon \int_{\mathscr{B}_R} |\varphi| \, dH_{n-1}$$

$$(2.20) \quad \int_{C_R^+} |Df| \, dz \leqq (1 + \varepsilon) \int_{\mathscr{B}_R} |\varphi| \, dH_{n-1}.$$

Proof: Let $\{\varphi_k\}$ be a sequence of C^∞-functions in $\mathscr{B}_R$, converging to φ in $L^1(\mathscr{B}_R)$. We can assume that $\varphi_0 = 0$ and

$$\| \varphi_k \| \leqq 2 \| \varphi \|$$

$$\sum_{k=0}^{\infty} \| \varphi_k - \varphi_{k+1} \| \leqq (1 + \varepsilon/2) \| \varphi \|$$

$\| \quad \|$ denoting the norm in $L^1(\mathscr{B}_R)$.

Let $\{t_k\}$ be a decreasing sequence, converging to zero. For $z = (x, t)$ we set

$$f(z) = f(x, t) = \begin{cases} 0 \text{ if } t > t_0 \\ \dfrac{t - t_{k+1}}{t_k - t_{k+1}} \varphi_k(x) + \dfrac{t_k - t}{t_k - t_{k+1}} \varphi_{k+1}(x) \text{ if } t_{k-1} \leqq t \leqq t_k. \end{cases}$$

We have for $t_{k+1} < t < t_k$

$$|D_i f| \leqq |D_i \varphi_k| + |D_i \varphi_{k+1}| \quad i = 1, 2, \ldots, n-1$$

$$|D_n f| = |\varphi_k - \varphi_{k+1}|(t_k - t_{k+1})^{-1}$$

and therefore

$$\int_{C_R^+} |Df| \, dz \leqq \sum_{k=0}^{\infty} \| \varphi_k - \varphi_{k+1} \| + \sum_{k=0}^{\infty} (\| D\varphi_k \| + \| D\varphi_{k+1} \|)(t_k - t_{k+1}).$$

Moreover:

$$\int\limits_{C_{\dot R}} |f|\,dz \le \sum_{k=0}^{\infty} (\|\varphi_k\| + \|\varphi_{k+1}\|)(t_k - t_{k+1}) \le 4\|\varphi\|\,t_0.$$

If we choose the sequence t_k in such a way that $4t_0 < \varepsilon$ and

$$t_k - t_{k+1} \le \frac{\varepsilon\|\varphi\|}{1 + \|D\varphi_k\| + \|D\varphi_{k+1}\|} 2^{-k-2}$$

we have immediately (2.19) and (2.20). The proof that the trace of f is actually φ proceeds exactly as in 2.4. $\square$

A simple argument based on a partition of unity (see 2.10) proves

2.16   Theorem: *Let Ω be a bounded open set with Lipschitz-continuous boundary $\partial\Omega$, and let $\varphi \in L^1(\partial\Omega)$. For every $\varepsilon > 0$ there exists a function $f \in W^{1,1}(\Omega)$ having trace φ on $\partial\Omega$ and such that*

(2.21) $\int\limits_{\Omega} |f|\,dx \le \varepsilon\|\varphi\|$

(2.22) $\int\limits_{\Omega} |Df|\,dx \le A\|\varphi\|$

with A depending on $\partial\Omega$, but independent of φ, f and ε.

2.17   Remark: It is easily seen from the construction of f that its support can be taken in an arbitrary neighborhood of $\partial\Omega$. Moreover, if $\partial\Omega$ is of class C^1, we may take $A = 1 + \varepsilon$.

3. The Reduced Boundary

Recalling that when considering functions in BV we are really considering equivalence classes of functions, we see that changing a function on a set of measure zero gives, as far as BV is concerned, the same function. In the same way when considering Caccioppoli sets the perimeter and other properties are unchanged if we make alterations of measure zero. In other words we are really concerned with equivalence classes of Caccioppoli sets.

3.1 Proposition: *If E is a Borel set, then there exists a Borel set $\tilde{E}$ equivalent to E (that is, differs only by a set of measure zero) and such that*

(3.1) $0 < |\bar{E} \cap B(x, \rho)| < \omega_n \rho^n$ *for all $x \in \partial \tilde{E}$ and all $\rho > 0$,*

where ω_n is the measure of the unit ball in $\mathbb{R}^n$.

Proof: Define

$$E_0 = \{x \in \mathbb{R}^n : \text{there exists } \rho > 0 \text{ with } |E \cap B(x, \rho)| = 0\}$$

and

$$E_1 = \{x \in \mathbb{R}^n : \text{there exists } \rho > 0 \text{ with } |E \cap B(x, \rho)| = |B(x, \rho)| = \omega_n \rho^n\}.$$

Suppose $x \in E_0$; then there exists $\rho > 0$ such that $|E \cap B(x, \rho)| = 0$. Now suppose $y \in B(x, \rho)$, and setting $\rho_0 = \rho - |x - y| > 0$ we have $B(y, \rho_0) \subseteq B(x, \rho)$ and $|B(y, \rho_0) \cap E| = 0$. Thus $B(x, \rho) \subseteq E_0$ and so E_0 is open. In the same way we can show that E_1 is open. We next establish that $|E_0 \cap E| = 0$.

For each $x \in E_0$ choose $\rho > 0$ such that $|E \cap B(x, \rho)| = 0$. Then $\{B(x, \rho) : x \in E_0\}$ forms an open covering of E_0 and so there exists a sequence $\{x_i\}$ in E_0 such that

$$E_0 \subseteq \bigcup_{i=1}^{\infty} B(x_i, \rho_i) \text{ and } |E \cap B(x_i, \rho_i)| = 0$$

and hence

$$|E \cap E_0| \leq |\bigcup_{i=1}^{\infty} (B(x_i, \rho_i) \cap E)| = 0.$$

Similarly we can show $|E_1 - E| = 0$.

Now set $\tilde{E} = (E \cup E_1) - E_0$. Then $\tilde{E}$ is equivalent to E and, since E_1 and E_0 are open, if $x \in \partial\tilde{E}$ then $x \notin E_1 \cup E_0$ and (3.1) must hold.   □

3.2   Remark: Since we are concerned only with equivalence classes of sets, by Proposition 3.1, we may assume that inequality (3.1) holds for every set we consider.

We now want to introduce a particular subset of the boundary of a Caccioppoli set known as the reduced boundary and denoted ∂^*E. This concept was introduced by De Giorgi [DG2] and it is of fundamental importance when considering the regularity of the boundary of minimizing sets.

We shall show later that, for a minimizing set E, the reduced boundary ∂^*E is analytic and we shall then be concerned with estimating the size of $\partial E - \partial^*E$ and in particular its Hausdorff dimension.

3.3   Definition: A point x belongs to the *reduced boundary*, ∂^*E, of a set E if

(3.2) $\int_{B(x,\rho)} |D\varphi_E| > 0$ for all $\rho > 0$,

(3.3)   the limit $v(x) = \lim_{\rho \to 0} v_\rho(x)$ exists

where

$$v_\rho(x) = \frac{\int_{B(x,\rho)} D\varphi_E}{\int_{B(x,\rho)} |D\varphi_E|}$$

and

(3.4) $|v(x)| = 1.$

Note that, from the theorem of Besicovitch on differentiation of measures

(2.9 of [FH2]), it follows that $v(x)$ exists and $|v(x)| = 1$ for $|D\varphi_E|$ – almost all $x \in \mathbb{R}^n$, and furthermore that

$$D\varphi_E = v|D\varphi_E|.$$

3.4   Examples: (i) Suppose ∂E is a C^1 hypersurface and $x \in \partial E$. Now, from Remark 2.13, setting $A = E$ and $f = \varphi_E$, we obtain that

$$D\varphi_E = v dH_{n-1} \text{ on } \partial E$$

where v is the unit inner normal to ∂E. Furthermore by (1.3), spt $D\varphi_E \subseteq \partial E$ and so

$$\int\limits_{B(x,\rho)} D\varphi_E = \int\limits_{B(x,\rho) \cap \partial E} v dH_{n-1}.$$

On the other hand from (1.1)

$$\int\limits_{B(x,\rho)} |D\varphi_E| = H_{n-1}(B(x,\rho) \cap \partial E)$$

and therefore

$$v_\rho(x) = \frac{\int\limits_{B(x,\rho) \cap \partial E} v dH_{n-1}}{H_{n-1}(B(x,\rho) \cap \partial E)}.$$

Hence, since v is continuous on ∂E,

$$\lim_{\rho \to 0} v_\rho(x) = v(x).$$

Thus x is in the reduced boundary. In conclusion, if ∂E is a C^1 hypersurface then $\partial^* E = \partial E$ and $v(x)$ is the unit inner normal vector to ∂E at x.

(ii) Let E be the unit square in $\mathbb{R}^2$, then conditions (3.2) and (3.3) are satisfied for each x in ∂E and (3.4) is satisfied except at the corner points where $|v| = \dfrac{1}{\sqrt{2}}$.

The properties of the sets we consider are unchanged by translation and rotation. For simplicity we shall often assume that the origin belongs to ∂E or that the x_1-axis is normal to ∂E and then by the appropriate rotation and translation we may prove similar results for any point in ∂E.

3.5   Lemma: *Suppose $E \subseteq \mathbb{R}^n$ is such that $0 \in \partial E$ and there exists a number $\bar{\rho} > 0$ such that for every $\rho < \bar{\rho}$*

$$\int\limits_{\overset{\circ}{B}_\rho} |D\varphi_E| > 0$$

$$|v_\rho(0)| = |\int\limits_{\overset{\circ}{B}_\rho} D\varphi_E| / \int\limits_{\overset{\circ}{B}_\rho} |D\varphi_E| \geq q > 0.$$

Then for every $\rho < \bar{\rho}$

(3.5) $\rho^{-n}|E \cap B_\rho| \geq C_1(n, q) > 0,$

(3.6) $\rho^{-n}|(\mathbb{R}^n - E) \cap B_\rho| \geq C_2(n, q) > 0,$

(3.7) $0 < C_3(n, q) \leq \rho^{1-n} \int\limits_{\overset{\circ}{B}_\rho} |D\varphi_E| \leq C_4(n, q),$

where $C_1(n, q)$, $C_2(n, q)$, $C_3(n, q)$ and $C_4(n, q)$ are constants depending only on n and q.

Proof: From (2.14), as in the proof of (2.16), we can prove that for almost all $\rho < \bar{\rho}$

$$0 = \int D\varphi_{E \cap B_\rho} = \int\limits_{\overset{\circ}{B}_\rho} D\varphi_E + \int\limits_{\partial B_\rho} \vec{n} \varphi_E dH_{n-1}$$

where $\vec{n}$ is the interior normal to ∂B_ρ. Thus it follows that

$$|\int\limits_{\overset{\circ}{B}_\rho} |D\varphi_E| \leq \int\limits_{\partial B_\rho} \varphi_E dH_{n-1} \, T = H_{n-1}(E \cap \partial B_\rho).$$

On the other hand, for every $\rho < \bar{\rho}$,

$$\int\limits_{\overset{\circ}{B}_\rho} |D\varphi_E| \leq \frac{1}{q} |\int\limits_{\overset{\circ}{B}_\rho} D\varphi_E|.$$

Hence

$$\int\limits_{\overset{\circ}{B}_\rho} |D\varphi_E| \leq \frac{1}{q} H_{n-1}(E \cap \partial B_\rho) \leq C_4 \rho^{n-1}$$

for almost all $\rho < \bar{\rho}$ and by the semicontinuity Theorem 1.9 it will in fact hold for every $\rho < \bar{\rho}$. This proves the second part of (3.7). By (2.16) in Remark 2.14, if we set

$$E_\rho = E \cap B_\rho$$

we have for almost all ρ

$$P(E_\rho) = P(E, B_\rho) + \int_{\partial B_\rho} \varphi_E dH_{n-1}.$$

But as above

$$P(E, B_\rho) \leqq \frac{1}{q} \int_{\partial B_\rho} \varphi_E dH_{n-1}$$

for almost all $\rho < \bar{\rho}$. Thus

$$P(E_\rho) \leqq C_5 \int_{\partial B_\rho} \varphi_E dH_{n-1}$$

and hence from the isoperimetric inequality (1.19)

$$|E_\rho|^{1-1/n} \leqq C_6 \int_{\partial B_\rho} \varphi_E dH_{n-1}.$$

On the other hand, if $g(\rho) = |E_\rho|$,

$$g(R) = \int_{B_R} \varphi_E dx = \int_0^R d\rho \int_{\partial B_\rho} \varphi_E dH_{n-1}$$

and so $g(\rho)$ is absolutely continuous with

$$g'(\rho) = \int_{\partial B_\rho} \varphi_E dH_{n-1}.$$

It then follows that

$$g'(\rho) \geqq \frac{1}{C_6} g(\rho)^{1-1/n}$$

and so

$$g(\rho) \geq \left(\frac{\rho}{nC_6} \right)^n,$$

proving (3.5).

In a similar way, noting that $\varphi_{\mathbb{R}^n - E} = 1 - \varphi_E$ we can prove (3.6). It remains only to prove the left hand side of (3.7). This follows from (3.5), (3.6) and the isoperimetric inequality (1.20).   □

Notice that, in particular, if $0 \in \partial^* E$ then there exists a number $\bar{\rho}$ such that the lemma holds.

3.6   Definition: For $z \in \partial^* E$, define the tangent hyperplane

$$T(z) = \{ x \in R^n : \langle v(z), x - z \rangle = 0 \}$$

and the sets

$$T^+(z) = \{ x \in \mathbb{R}^n : \langle v(z), x - z \rangle > 0 \},$$

$$T^-(z) = \{ x \in \mathbb{R}^n : \langle v(z), x - z \rangle < 0 \}.$$

3.7   Theorem: *Let* $0 \in \partial^* E$ *and for* $t > 0$ *define*

$$E_t = \{ x \in \mathbb{R}^n : tx \in E \}.$$

Then as $t \to 0^+$ *the set* E_t *converges to* $T^+(0)$ *and moreover, for every set* A *such that* $H_{n-1}(\partial A \cap T(0)) = 0,$

(3.8) $\lim\limits_{t \to 0} \int\limits_A |D\varphi_{E_t}| = \int\limits_A |D\varphi_{T^+}| = H_{n-1}(A \cap T(0)).$

Proof: By the reasoning mentioned before Lemma 3.5 we can suppose that

$$v_1(0) = -1, \quad v_2(0) = \ldots = v_n(0) = 0$$

so that

$$T^+ = T^+(0) = \{ x : x_1 < 0 \}.$$

It is sufficient to show that every sequence $\{ t_j \} \to 0$ has a subsequence $\{ s_j \}$ such that $E_{s_j} \to T^+$, and if we denote $E_j = E_{s_j}$, then

$$\lim_{j \to \infty} \int_A |D\varphi_{E_j}| = \int_A |D\varphi_{T^+}|.$$

Now for every $\rho > 0$, by a simple change of variables,

$$(3.9) \quad \int_{B_\rho} D\varphi_{E_t} = t^{1-n} \int_{B_{t\rho}} D\varphi_E,$$

$$(3.10) \quad \int_{B_\rho} |D\varphi_{E_t}| = t^{1-n} \int_{B_{t\rho}} |D\varphi_E|$$

and therefore by the definition of $v(0)$

$$(3.11) \quad \lim_{t \to 0} \frac{\int_{B_\rho} D_1\varphi_{E_t}}{\int_{B_\rho} |D\varphi_{E_t}|} = v_1(0) = -1$$

$$(3.12) \quad \lim_{t \to 0} \frac{\int_{B_\rho} D_i\varphi_{E_t}}{\int_{B_\rho} |D\varphi_{E_t}|} = 0 \quad i = 2, 3, \ldots, n.$$

Moreover from (3.10) and (3.7) of Lemma 3.5, we obtain

$$(3.13) \quad \limsup_{t \to 0} \int_{B_\rho} |D\varphi_{E_t}| < \infty.$$

Now suppose we have a sequence $\{t_j\}$ converging to 0; then, by (3.13) and the compactness Theorem 1.19, there will exist a subsequence $\{s_j\}$ and a Caccioppoli set C such that if $E_j = E_{s_j}$ then φ_{E_j} converges in $L^1_{loc}(\mathbb{R}^n)$ to φ_C. Moreover by the De La Vallée Poussin Theorem (see Appendix A) we may assume that

$$(3.14) \quad \lim_{j \to \infty} \int_{B_\rho} D\varphi_{E_j} = \int_{B_\rho} D\varphi_C$$

for almost all ρ (in particular for those ρ for which $\int_{\partial B_\rho} |D\varphi_C| = 0$). Now from (3.11) and (3.14)

$$\lim_{j \to \infty} \int_{B_\rho} |D\varphi_{E_j}| = -\lim_{j \to \infty} \int_{B_\rho} D_1\varphi_{E_j} = -\int_{B_\rho} D_1\varphi_C.$$

Thus by semicontinuity

(3.15) $\int_{B_\rho} |D\varphi_C| \leqq -\int_{B_\rho} D_1\varphi_C$

and, by definition of $D_1\varphi_C$, we get equality in (3.15). By differentiation of measures we obtain

$$|D\varphi_C| + D_1\varphi_C = 0$$

and hence

$$D_i\varphi_C = 0 \quad i = 2, \ldots, n.$$

Therefore φ_C depends only on x_1 and furthermore it is a non-increasing function of x_1. This implies that there exists a $\lambda \in \mathbb{R}$ such that

$$C = \{x \in \mathbb{R}^n : x_1 < \lambda\}.$$

We now show that $\lambda = 0$. Suppose that $\lambda < 0$; then since $\varphi_{E_j} \to \varphi_{E_C}$ in $L^1_{loc}(\mathbb{R}^n)$, we have

$$0 = |C \cap B_{|\lambda|}| = \lim_{j \to \infty} |E_j \cap B_{|\lambda|}| = \lim_{j \to \infty} s_j^{-n}|E \cap B_{|\lambda|s_j}|$$

which contradicts (3.5). Similarly if $\lambda > 0$ we get a contradiction to (3.6). Thus $\lambda = 0$, $C = T^+$ and

(3.16) $\int_{B_\rho} |D\varphi_{T^+}| = \lim_{j \to \infty} \int_{B_\rho} |D\varphi_{E_j}|$

for almost every ρ.

Finally let A be an open set such that

$$\int_{\partial A} |D\varphi_{T^+}| = H_{n-1}(T \cap \partial A) = 0.$$

Choose ρ such that $A \subseteq B_\rho$ and (3.16) holds; then by Proposition 1.13 we obtain equation (3.8). □

The above theorem shows that, at least in some sense, $T(0)$ is indeed a tangent plane to the surface ∂E at 0. This is made a little clearer in the next theorem.

3.8   Theorem: *Suppose $E \subseteq \mathbb{R}^n$ and $0 \in \partial^* E$. For $\rho, \varepsilon > 0$ define*

$$S_{\rho,\varepsilon} = B_\rho \cap \{x \in \mathbb{R}^n : |\langle v(0), x \rangle| < \varepsilon \rho\}.$$

Then

(3.17) $\displaystyle \lim_{\rho \to 0} \rho^{1-n} \int_{S_{\rho,\varepsilon}} |D\varphi_E| = \omega_{n-1},$

(3.18) $\displaystyle \lim_{\rho \to 0} \rho^{-n} |E \cap B_\rho \cap T^-| = 0,$

(3.19) $\displaystyle \lim_{\rho \to 0} \rho^{-n} |(B_\rho - E) \cap T^+| = 0,$

where ω_{n-1} is the measure of the unit ball in $\mathbb{R}^{n-1}$,

Proof: From (3.10)

$$\rho^{1-n} \int_{B_\rho} |D\varphi_E| = \int_{B_1} |D\varphi_{E_\rho}|$$

and similarly

$$\rho^{1-n} \int_{S_{\rho,\varepsilon}} |D\varphi_E| = \int_{S_{1,\varepsilon}} |D\varphi_{E_\rho}|.$$

By (3.8),

$$\lim_{\rho \to 0} \int_{S_{1,\varepsilon}} |D\varphi_{E_\rho}| = H_{n-1}(S_{1,\varepsilon} \cap T(0)) = \omega_{n-1}.$$

and so (3.17) holds. To prove (3.18), first note that

$$\rho^{-n} |E \cap B_\rho \cap T^-| = |E_\rho \cap B_1 \cap T^-|.$$

Now by Theorem 3.7

$$\lim_{\rho \to 0} |E_\rho \cap B_1 \cap T^-| = |T^+ \cap B_1 \cap T^-| = 0$$

and (3.18) follows. Similarly we may prove (3.19).   □

3.9   Remark: Noting that $|B_\rho| = \rho^n \omega_n$, (3.18) says that the ratio of the measure of $(E \cap T^-) \cap B_\rho$ to the measure of B_ρ goes to zero as $\rho \to 0$. In other words, in small enough balls most of E lies in T^+. Similarly (3.19) says that in small enough balls most of $\mathbb{R}^n - E$ lies in T^-, so that in small enough balls B_ρ, the hyperplane T splits B_ρ into two parts which nearly correspond to E and $\mathbb{R}^n - E$.

Consider (3.17) in the case where ∂E is smooth. In this case $v(0)$ will be the normal to ∂E at 0 and $S_{\rho,\varepsilon}$ will be a strip of width $2\varepsilon\rho$ and centre $T(0)$ lying in the ball B_ρ.

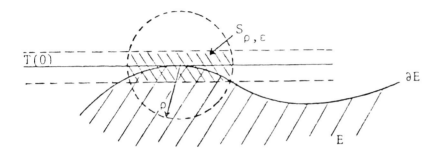

Since ∂E is smooth,

$$\int_{S_{\rho,\varepsilon}} |D\varphi_E| = H_{n-1}(\partial E \cap S_{\rho,\varepsilon})$$

and also $H_{n-1}(T(0) \cap B_\rho) = \rho^{n-1}\omega_{n-1}$. We see then that (3.17) says that by taking ρ sufficiently small we may ensure that ∂E mostly lies in the strip $S_{\rho,\varepsilon}$ and nearly corresponds to the hyperplane $T(0)$.

4. Regularity of the Reduced Boundary

In this chapter we establish some further important properties of the reduced boundary ∂^*E. We show first that ∂^*E may be written, up to a set of $|D\varphi_E|$-measure zero, as a countable union of C^1 hypersurfaces [DG2]. Furthermore we show that ∂^*E is dense in ∂E and that

$$(4.1) \quad \int_\Omega |D\varphi_E| = H_{n-1}(\partial^*E \cap \Omega)$$

for every open set Ω, so that $|D\varphi_E|$ is just $(n-1)$-dimensional Hausdorff measure restricted to ∂^*E.

To prove (4.1) we first show that if $B \subseteq \partial^*E$ then

$$(4.2) \quad \int_B |D\varphi_E| = H_{n-1}(B)$$

and we begin this by estimating the ratio of the two terms in (4.2).

4.1 Lemma: *Let E be a Caccioppoli set in $\mathbb{R}^n$ and let $B \subseteq \partial^*E$, then*

$$H_{n-1}(B) \leqq 2.3^{n-1} \int_B |D\varphi_E|.$$

Proof: Suppose $\varepsilon, \eta > 0$. As $|D\varphi_E|$ is a Radon measure there exists an open set A such that $B \subseteq A$ and

$$\int_A |D\varphi_E| \leqq \int_B |D\varphi_E| + \eta.$$

From Theorem 3.8, equation (3.17), for each $x \in B$ there exists a number $\rho = \rho(x) > 0$ such that $B(x, \rho) \subseteq A$, $\rho < \varepsilon$ and

$$\int_{B(x,\rho)} |D\varphi_E| \geqq \tfrac{1}{2}\omega_{n-1}\rho^{n-1}.$$

Now by Lemma 2.2, we can select a sequence $\{x_i\} \subseteq B$ such that, if $\rho_i = \rho(x_i)$,

$$B \subseteq \bigcup_{i=1}^{\infty} B(x_i, 3\rho_i) \quad \text{and} \quad B(x_i, \rho_i) \cap B(x_j, \rho_j) = \varnothing$$

for $i \neq j$.

Using the facts that $B(x_i, \rho_i) \subseteq A$ and are disjoint, we obtain

$$\sum_{i=1}^{\infty} (3\rho_i)^{n-1} \leq \frac{2.3^{n-1}}{\omega_{n-1}} \sum_{i=1}^{\infty} \int_{B(x_i, \rho_i)} |D\varphi_E| \leq$$

$$\leq \frac{2.3^{n-1}}{\omega_{n-1}} \int_A |D\varphi_E| \leq$$

$$\leq \frac{2.3^{n-1}}{\omega_{n-1}} \left\{ \int_B |D\varphi_E| + \eta \right\}.$$

So, by the fact that ε, η are arbitrary and by the definition of Hausdorff measure,

$$H_{n-1}(B) \leq 2.3^{n-1} \int_B |D\varphi_E|. \qquad \square$$

4.2   Definition: Γ_{n-1} is the class of all sets $H \subseteq \mathbb{R}^n$ such that there exists an open set A containing $\bar{H}$ and a C^1 function $f: A \to \mathbb{R}$ such that

$$f(x) = 0 \quad \text{and} \quad Df(x) \neq 0 \quad \text{for } x \in \bar{H}.$$

The following theorem gives a useful means of determining when a set is in Γ_{n-1}.

4.3   Theorem: *Let C be a compact set and suppose that there exists a vector valued continuous function $v: C \to \mathbb{R}^n$ such that $v \neq 0$ and*

$$(4.3) \qquad \lim_{|x-y| \to 0} \, <v(x), x-y> |x-y|^{-1} = 0$$

uniformly for $x, y \in C$. Then $C \in \Gamma_{n-1}$.

Proof: From the Whitney Extension Theorem [WH], [MB], there exists a function $f: \mathbb{R}^n \to \mathbb{R}$ which satisfies $f \in C^1$, $f = 0$ on C and $Df = v$ on C. Then as $v \neq 0$ we have $C \in \Gamma_{n-1}$. $\qquad \square$

4.4 Theorem [DG2]: *If E is a Caccioppoli set, then*

(4.4) $\partial^* E = \bigcup_{i=1}^{\infty} C_i \cup N$

where $\int_N |D\varphi_E| = 0$ *and each* C_i *is compact and belongs to* Γ_{n-1}. *Moreover, for*

every set $B \subseteq \partial^* E$

(4.5) $\int_B |D\varphi_E| = H_{n-1}(B),$

for every open set $\Omega \subseteq \mathbb{R}^n$

(4.6) $P(E, \Omega) = \int_\Omega |D\varphi_E| = H_{n-1}(\partial^* E \cap \Omega)$

and finally

(4.7) $\overline{\partial^* E} = \partial E.$

Proof: For each $x \in \partial^* E$ we know that (3.18) and (3.19) hold after suitable translation. Then for each integer i, by Egoroff's theorem [HP], [MME], we may choose a measurable set F_i such that $|D\varphi_E|(\partial^* E - F_i) < \dfrac{1}{2i}$ and the convergence in (3.18) and (3.19) is uniform. Further by Lusin's theorem [HP], [MME], we may choose a compact set C_i such that $|D\varphi_E|(F_i - C_i) < \dfrac{1}{2i}$ and the function v restricted to C_i is continuous. This construction gives (4.4) and it remains only to show that $C_i \in \Gamma_{n-1}$.

Let us consider one such C_i, for example C_1. By our choice of C_1, for every ε with $0 < \varepsilon < 1$ there exists a σ such that $0 < \sigma < 1$ and, if $\rho < 2\sigma$ and $z \in C_1$, then

$$|E \cap B(z, \rho) \cap T^-(z)| < \tfrac{1}{4}\varepsilon^n \omega_n \rho^n 2^{-n}$$

and

$$|E \cap B(z, \rho) \cap T^+(z)| > \frac{\omega_n \rho^n}{2} - \varepsilon^n \frac{\omega_n \rho^n}{4} 2^{-n} = \frac{\omega_n \rho^n}{2}\left(1 - \frac{\varepsilon^n}{2^{n+1}}\right).$$

We shall prove that for every $x, y \in C_1$ such that $|x - y| < \sigma$ we have

$$|\langle v(x), x - y \rangle| \ |x - y|^{-1} \leq \varepsilon.$$

Hence as ε is arbitrary we may apply Theorem 4.3 to obtain $C_1 \in \Gamma_{n-1}$. Suppose first that

$$\langle v(x), y - x \rangle < -\varepsilon |x - y|.$$

Since $\varepsilon < 1$, we have

$$(4.8) \quad B(y, \varepsilon|x - y|) \subseteq T^-(x) \cap B(x, 2|x - y|).$$

On the other hand, since $|x - y| < \sigma$,

$$(4.9) \quad |E \cap B(x, 2|x - y|) \cap T^-(x)| < \frac{\varepsilon^n \omega_n |x - y|^n}{4}$$

and

$$(4.10) \quad |E \cap B(y, \varepsilon|x - y|)| \geq |E \cap B(y, \varepsilon|x - y|) \cap T^-(y)|$$

$$\geq \frac{\omega_n \varepsilon^n |x - y|^n}{2}\left(1 - \frac{\varepsilon^n}{2^{1+n}}\right) > \frac{\omega_n \varepsilon^n |x - y|^n}{4}.$$

The inequalities (4.9) and (4.10) contradict the inclusion of (4.8).

In a similar way we can prove also that

$$\langle v(x), y - x \rangle < \varepsilon |x - y|$$

and so (4.3) must hold.

Also using the above techniques we may show that $C_i \in \Gamma_{n-1}$ for $i = 2, 3, \ldots$. To prove (4.5), observe that by Lemma 4.1

$$H_{n-1}(B - C_i) \leq 2.3^{n-1} |D\varphi_E|(B - C_i) < \frac{2.3^{n-1}}{i}$$

and so it is sufficient to prove (4.5) for $B \cap C_i$ or in other words for $B \in \Gamma_{n-1}$.

If $B \in \Gamma_{n-1}$, then by Definition 4.2 there exists an open set A containing $\bar{B}$ and a C^1 function $f: A \to \mathbb{R}^n$ such that

$$f = 0 \quad \text{and} \quad Df \neq 0 \text{ in } \bar{B}.$$

Since f is C^1, we may as well assume that $Df \neq 0$ in A so that

$$V = \{x \in A, f(x) = 0\}$$

is a regular hypersurface and $\bar{B} \subset V$. Let γ denote the measure $H_{n-1} \, \llcorner \, V$, that is $(n-1)$-dimensional Hausdorff measure restricted to V. Then because $\bar{B} \subseteq A$ and A is open, by properties of Hausdorff measure

$$\lim_{\rho \to 0} \rho^{1-n} \gamma(B(x, \rho)) = \omega_{n-1} \quad \text{for each } x \in B$$

and hence as $x \in B \subset \partial^* E$, from (3.8) we obtain

$$\lim_{\rho \to 0} \frac{\gamma(B(x, \rho))}{\int_{B(x, \rho)} |D\varphi_E|} = 1 \quad \text{for each } x \in B.$$

This implies, by the differentiation of measures, that

$$H_{n-1}(B) = \int_B |D\varphi_E|.$$

In particular

$$H_{n-1}(\partial^* E \cap \Omega) = \int_{\partial^* E \cap \Omega} |D\varphi_E|.$$

On the other hand, by the Besicovitch theorem [FH2] the vector $v(x)$ exists and $|v(x)| = 1$, $|D\varphi_E|$-almost everywhere in ∂E. Thus the set $\partial E - \partial^* E$ has $|D\varphi_E|$ measure zero and using (1.3)

$$P(E, \Omega) = \int_\Omega |D\varphi_E| = \int_{\Omega \cap \partial E} |D\varphi_E| = \int_{\Omega \cap \partial^* E} |D\varphi_E|.$$

Finally let A be an open set such that

$$\partial^* E \cap A = \varnothing;$$

then by (4.6)

$$\int_A |D\varphi_E| = 0.$$

Hence φ_E is constant in A and so $\partial E \cap A = \varnothing$. Therefore $\overline{\partial^* E} = \partial E$. □

Suppose α is a unit vector in $\mathbb{R}^n$ with $\alpha = (\alpha_1, \ldots, \alpha_n)$. Then we denote

$$D_\alpha = \sum_{i=1}^{n} \alpha_i D_i$$

4.5   Lemma: *Let E be a Caccioppoli set in Ω, let $z \in \Omega$ and $\rho > 0$ and suppose that there exists a number $\tau > 0$ such that, for all t with $0 < t < \tau$, the ball $B(z + t\alpha, \rho)$ is strictly contained in Ω. Then*

$$(4.11) \quad |E \cap B(z + \tau\alpha, \rho)| - |E \cap B(z, \rho)| = \int_0^\tau dt \int_{B(z + t\alpha, \rho)} D_\alpha \varphi_E.$$

Proof: Suppose $g \in C_0^\infty(\Omega)$ and spt $g(x - t\alpha) \subset\subset \Omega$ for every $t < \tau$; then

$$\int_E [g(x - \tau\alpha) - g(x)] dx = -\int_E dx \int_0^\tau D_\alpha g(x - t\alpha) dt =$$

$$= -\int_0^\tau dt \int \varphi_E D_\alpha g(x - t\alpha) dx =$$

$$= \int_0^\tau dt \int g(x - t\alpha) D_\alpha \varphi_E.$$

Now choose functions $g_k \in C_0^\infty(\Omega)$ such that $0 \leq g_k \leq 1$, $g_k = 1$ in $B\left(z, \rho - \frac{1}{k}\right)$ and spt $g_k \subseteq B(z, \rho)$. (This will be possible for large enough k.) Writing the above equation for each g_k and then passing to the limit we obtain

$$|E \cap B(z + t\alpha, \rho)| - |E \cap B(z, \rho)| = \int_0^\tau dt \int_{B(z + t\alpha, \rho)} D_\alpha \varphi_E$$

and the conclusion follows.   □

4.6   Lemma: *Let E be a Caccioppoli set in Ω and suppose that there exists a vector α in $\mathbb{R}^n$ and a positive real number p such that*

$$(4.12) \quad v(x) \cdot \alpha = \lim_{\rho \to 0} \frac{\int_{B(x,\rho)} D_\alpha \varphi_E}{\int_{B(x,\rho)} |D\varphi_E|} \geq p > 0$$

for $|D\varphi_E|$-almost all x in Ω. Suppose $z \in \partial E \cap \Omega$ and $k > 0$ is such that the segment $[z, z + k\alpha] \subseteq \Omega$. Then $z + k\alpha$ is interior to E.

Proof: Suppose there exists a $z \in \partial E \subseteq \Omega$ and a $k > 0$ such that $[z, z + k\alpha] \subseteq \Omega$ and $z + k\alpha$ is not interior to E. We show first that $[z, z + k\alpha] \subseteq \partial E$.

Suppose there exists a point $z + \tau\alpha \in \Omega - \bar{E}$; then choose $\rho > 0$ such that $B(z + \tau\alpha, \rho) \subseteq \Omega - \bar{E}$. Then from Lemma 4.5 and inequality (4.12)

$$0 \le \int_0^\tau dt \int_{B(z + t\alpha, \rho)} D_\alpha \varphi_E = |E \cap B(z + \tau\alpha, \rho)| - |E \cap B(z, \rho)| =$$

$$= -|E \cap B(z, \rho)| < 0$$

which gives a contradiction.

Alternatively, suppose there exists a point $z + \tau\alpha \in E - \partial E$ and that $z + k\alpha \in \partial E$. Choose $\rho > 0$ such that $B(z + \tau\alpha, \rho) \cap E$. From Lemma 4.5

$$|E \cap B(z + k\alpha, \rho)| - |E \cap B(z + \tau\alpha, \rho)| = \int_\tau^k dt \int_{B(z + t\alpha, \rho)} D_\alpha \varphi_E \ge 0.$$

But by the choice of ρ, $|E \cap B(z + \tau\alpha, \rho)| = \omega_n \rho^n$ and by (3.1), since $z + k\alpha \in \partial E$, $|E \cap B(z + k\alpha, \rho)| < \omega_n \rho^n$ and again we have a contradiction. Thus $[z, z + k\alpha] \subseteq \partial E$.

We now show that this also gives a contradiction and the lemma follows. Choose ρ_0 so that $B(z + t\alpha, \rho) \subseteq \Omega$ for each $\rho \le \rho_0$ and each $0 < t < k$. Then, by definition of v and $D_\alpha \varphi_E$ and by Theorem 4.4,

$$\int_{B(z + t\alpha, \rho)} D_\alpha \varphi_E = \int_{B(z + t\alpha, \rho)} v \cdot \alpha |D\varphi_E| \ge p \int_{B(z + ta, \rho)} |D\varphi_E|, \quad \text{by (4.12)}$$

Lemma 3.5 yields

$$\int_{B(z + t\alpha, \rho)} |D\varphi_E| > C_3 \rho^{n-1}$$

for each $0 < t < k$ and each $0 < \rho \le \rho_0$. Now from (4.11) and (4.12)

$$|E \cap B(z + k\alpha, \rho)| - |E \cap B(z, \rho)| = \int_0^k dt \int_{B(z + t\alpha, \rho)} D_\alpha \varphi_E \ge kp C_3 \rho^{n-1}$$

But the left hand side is bounded above by $\omega_n \rho^n$ and so we obtain a contradiction as $\rho \to 0$. $\square$

4.7 **Remark:** If the hypothesis of Lemma 4.6 held with $k < 0$ instead of $k > 0$, then the same argument would show that $z + k\alpha$ lies in the interior of $\mathbb{R}^n - E$.

We already know that $|v(x)|$ is bounded $|D\varphi_E|$-almost everywhere, in fact $|v(x)| = 1$ $|D\varphi_E|$-almost everywhere. Now we wish to show that if the direction of $v(x)$ does not vary too much then the set E has Lipschitz continuous boundary. By rotating if necessary we shall consider only the case where $v(x)$ is close to the x_n-axis.

4.8 **Theorem:** *Let Ω be a convex open set in $\mathbb{R}^n$ and let E be a Caccioppoli set in Ω. Suppose that*

$$v_n(x) = \lim_{\rho \to 0} \frac{\int\limits_{B(x,\rho)} D\varphi_E}{\int\limits_{B(x,\rho)} |D\varphi_E|} \geq q > 0$$

for some fixed constant q and $|D\varphi_E|$-almost all x in Ω. Then there exists an open set $A \subseteq \mathbb{R}^{n-1}$ and a function $f : A \to \mathbb{R}$ such that

(4.13) $\partial E \cap \Omega = \{(y, t) : y \in A, \ t = f(y)\}$

and moreover

(4.14) $|f(y) - f(y')| \leq q^{-1}\sqrt{1 - q^2}\,|y - y'|$

for all $y, y' \in A$.

Proof: Let $\alpha = (\alpha_1, \ldots, \alpha_n)$ be a unit vector with $\alpha_n > 0$; then

$$D_\alpha \varphi_E = \alpha_n D_n \varphi_E + \sum_{i=1}^{n-1} \alpha_i D_i \varphi_E \geq \left\{ \alpha_n q - \sqrt{(1 - \alpha_n^2)(1 - q^2)} \right\} |D\varphi_E|.$$

Hence, if $\alpha_n > \sqrt{1 - q^2}$ we can conclude by Lemma 4.6 that, for every $z \in \partial E \cap \Omega$, points in Ω of the form $z + t\alpha$ with $t > 0$ belong to the interior of E and those of the form $z - t\alpha$ belong to the interior of $\mathbb{R}^n - E$. In other words, for every $z \in \partial E \cap \Omega$, the intersection of Ω with the cone

$$C = \left\{ x \in \mathbb{R}^n : (x_n - z_n) > q^{-1}\sqrt{1 - q^2} \left[\sum_{i=1}^{n-1} (x_i - z_i)^2 \right]^{1/2} \right\}$$

is interior to E, and the intersection of Ω with

$$C' = \left\{ x \in \mathbb{R}^n : (x_n - z_n) < - q^{-1}\sqrt{1 - q^2} \left[\sum_{i=1}^{n-1} (x_i - z_i)^2 \right]^{1/2} \right\}$$

is interior to $\mathbb{R}^n - E$. From these arguments (4.13) and (4.14) follow immediately.   □

In the special case $\Omega = B_\rho$ we can make Theorem 4.8 a little more precise at least if we suppose that q is close to 1 and $\partial E \cap B(0, (1 - q)\rho)$ is nonempty. In other words we assume that $v(x)$ is always close to the x_n-axis and that ∂E passes close to the centre of the ball B_ρ. For simplicity we write $q = 1 - \sigma$ and, for $r > 0$, as usual we denote

$$\mathscr{B}_r = \{ y \in \mathbb{R}^{n-1} : |y| < r \}.$$

4.9   Proposition: *Let E be a Caccioppoli set in B_ρ such that*

$$v_n(x) \geq 1 - \sigma \quad \text{for } |D\varphi_E|\text{-almost all } x \text{ in } B_\rho,$$

where

$$\varepsilon = \frac{2\sqrt{2\sigma}}{1 - \sigma} < \frac{1}{2},$$

and suppose $\partial E \cap B_{\sigma\rho} \neq \varnothing$. Then, if f and A are as in Theorem 4.8,

(4.15) $A \supseteq \bar{\mathscr{B}}_{(1 - \varepsilon)\rho}$

and

(4.16) $|f(y)| \leq \varepsilon\rho, \quad |Df(y)| \leq \varepsilon/2$

for all y in $\bar{\mathscr{B}}_{(1-\varepsilon)\rho}$.

Proof: We have

$$q^{-1}\sqrt{1-q^2} = (1-\sigma)^{-1}\sqrt{2\sigma - \sigma^2} \leq \frac{\sqrt{2\sigma}}{1-\sigma} = \varepsilon/2.$$

and hence, from (4.14), $|Df(y)| \leq \varepsilon/2$.

Now let $z = (\eta, f(\eta)) \in \partial E \cap B_{\sigma\rho}$; then we have $|\eta| < \sigma\rho$ and $|f(\eta)| < \sigma\rho$ and so, if $y \in A$,

$$|f(y)| \leq |f(\eta)| + |f(y) - f(\eta)| < \sigma\rho + \frac{\varepsilon}{2}(|y| + |\eta|) <$$

$$< \sigma\rho + \frac{\varepsilon}{2}(\rho + \sigma\rho) = \rho\left(\sigma + \frac{\varepsilon}{2} + \frac{\sigma\varepsilon}{2}\right).$$

But $\sigma\varepsilon = \varepsilon - 2\sqrt{2\sigma}$, so that

$$|f(y)| \leq \rho(\varepsilon + \sigma - \sqrt{2\sigma}) \leq \varepsilon\rho.$$

Thus we need only to show (4.15). Obviously, by the condition on σ,

$$\sigma + \varepsilon < 1$$

and so $\mathscr{B}_{\sigma\rho} \subseteq \mathscr{B}_{(1-\varepsilon)\rho}$. Now if η is the point in A determined above, then $\eta \in A \cap \mathscr{B}_{\sigma\rho}$, and so $A \cap \mathscr{B}_{(1-\varepsilon)\rho}$ is nonempty. We prove (4.15) by showing that $\partial A \cap \bar{\mathscr{B}}_{(1-\varepsilon)\rho} = \varnothing$.

Suppose $y \in \partial A$; then obviously by (4.13) and (4.14) we must have $(y, f(y)) \in \partial B_\rho$ and so $|y| + |f(y)| \geq \rho$. But by the estimate above

$$|f(y)| \leq \rho(\varepsilon + \sigma - \sqrt{2\sigma})$$

and so

$$|y| \geq \rho(1 - \varepsilon + \sqrt{2\sigma} - \sigma) > (1 - \varepsilon)\rho. \qquad \square$$

4.10   Remark: It is obvious, from the proof of Theorem 4.8, that since $\partial E \cap \Omega = \{(y, t) : y \in A, \ t = f(y)\}$ then if E is open

$$E \cap \Omega = \Omega \cap \{(y, t): y \in A, t > f(y)\}.$$

Moreover, as f is Lipschitz continuous, it is differentiable almost everywhere in A and so, as in Example 3.4 (i), $v(x)$ will be the normal to the surface at almost all points and

$$v_i(x) = \frac{D_i f(y)}{\sqrt{1 + |Df(y)|^2}} \qquad i = 1, \ldots, n-1$$

$$v_n(x) = \frac{1}{\sqrt{1 + |Df(y)|^2}}$$

where $x = (y, f(y))$.

4.11 Theorem: *Suppose E is a Caccioppoli set in Ω such that $v(x)$ exists for every $x \in \partial E \cap \Omega$ and is continuous. Then $\partial E \cap \Omega$ is a C^1 hypersurface.*

Proof: We know that $|v(x)| = 1$ on $\partial^* E$ and that $\overline{\partial^* E} = \partial E$. Hence, as $v(x)$ is continuous, we have $|v(x)| = 1$ everywhere on ∂E. From Theorem 4.8 it follows that, for every $z \in \partial E \cap \Omega$, there exists a ball B such that $\partial E \cap B$ has a representation as a Lipschitz continuous function f. Now by Remark 4.10 we have for almost all y

$$D_i f(y) = v_i(x)/v_n(x),$$

where $x = (y, f(y))$. Hence the derivatives of f will coincide almost everywhere with continuous functions and so f must be C^1 itself. $\square$

5. Some Inequalities

In the last two chapters we have been concerned with the regularity of the reduced boundary of arbitrary Caccioppoli sets. We now want to turn our attention to the regularity of sets with minimal perimeter. However before doing this it is convenient to introduce some notation which will simplify things in later chapters and also to state and prove some preliminary lemmas ([DG5], [MM3]). These lemmas are mainly technical without introducing many new ideas and could easily be left until the appropriate step in the regularity proof.

In the first chapter we proved the existence of Caccioppoli sets with minimal perimeter. So generally in what follows we shall have such a minimizing set and then examine the regularity. In doing this it will often be necessary to approximate the minimal set and so it is appropriate to introduce a measure of just how close a set is to being minimal.

5.1 Definition: If $f \in BV(\Omega)$ then denote

$$v(f, \Omega) = \inf\{\int_\Omega |Dg| : g \in BV(\Omega), \operatorname{spt}(g - f) \subset \Omega\}$$

$$\psi(f, \Omega) = \int_\Omega |Df| - v(f, \Omega).$$

If $\Omega = B_\rho$, then we write $v(f, \rho)$ in place of $v(f, B_\rho)$ and $\psi(f, \rho)$ in place of $\psi(f, B_\rho)$. Moreover, if f is the characteristic function of some set E with finite perimeter, we shall write $v(E, \Omega)$ and $\psi(E, \Omega)$ instead of $v(\varphi_E, \Omega)$, $\psi(\varphi_E, \Omega)$.

5.2 Remark: If E is a minimal set in Ω, then $\psi(E, \Omega) = 0$.

We now turn our attention to some technical lemmas.

5.3 Lemma: *Let* $f \in BV(B_R)$ *and let* $0 < \rho < r < R$; *then*

$$(5.1) \qquad \int_{\partial B_1} |f^-(rx) - f^-(\rho x)| dH_{n-1} \leq \int_{B_r - B_\rho} |\langle \frac{x}{|x|^n}, Df \rangle|$$

Proof: Note that if $g \in C^1(A; \mathbb{R}^n)$ then $\int_A |\langle g, Df \rangle|$ is the total variation in A of the measure $\langle g, Df \rangle$, that is

$$\int_A |\langle g, Df \rangle| = \sup\{\int f \operatorname{div}(\mu g) dx : \mu \in C_0^1(A), |\mu| \leq 1\}.$$

Now let $g(x) = x|x|^{-n}$, h be any C^1 function and α be defined by

$$\alpha(x) = h(x|x|^{-1}).$$

Then $\operatorname{div}(\alpha g) = 0$ in $\mathbb{R}^n - \{0\}$ and so from (2.14)

$$\int_{B_r - B_\rho} \alpha \langle g, Df \rangle = \int_{\partial B_r} \alpha f^- \langle g, \frac{x}{|x|} \rangle dH_{n-1} - \int_{\partial B_\rho} \alpha f^+ \langle g, \frac{x}{|x|} \rangle dH_{n-1} =$$

$$= r^{1-n} \int_{\partial B_r} \alpha f^- dH_{n-1} - \rho^{1-n} \int_{\partial B_\rho} \alpha f^+ dH_{n-1} =$$

$$= \int_{\partial B_1} h(x)[f^-(rx) - f^+(\rho x)] dH_{n-1},$$

where in the last step we have taken into account the definition of $\alpha(x)$. Now, if we restrict h so that $|h(x)| \leq 1$ and hence $|\alpha(x)| \leq 1$, then by the definition of $\int_A |\langle g, Df \rangle|$:

$$\int_{\partial B_1} h(x)[f^-(rx) - f^+(\rho x)] dH_{n-1} \leq \int_{B_r - B_\rho} |\langle g, Df \rangle|$$

for any function h such that h is C^1 and $|h| \leq 1$.

Now for almost all $\rho < r$ we have $\int_{\partial B_\rho} |Df| = 0$ and $f^+ = f^- = f$ (see Remark 2.13), so that

$$(5.2) \quad \int_{\partial B_1} h(x)[f^-(rx) - f^-(\rho x)] dH_{n-1} \leq \int_{B_r - B_\rho} |\langle g, Df \rangle|$$

for almost all $\rho < r$. Thus if we take any $\rho < r$, we can choose a sequence $\{\rho_j\}$ such that $\rho_j \to \rho$, (5.2) holds for each ρ_j and $f^-(\rho_j x) \to f^-(\rho x)$ in $L^1(\partial B_1)$. Taking the limit as $j \to \infty$ we obtain (5.2) for every $\rho < r$. Now on taking the supremum over all h with $|h| \leq 1$ we arrive at inequality (5.1).   □

5.4   Remark: In a similar way we could obtain the inequality

$$\int_{\partial B_1} |f^+(rx) - f^+(\rho x)| \, dH_{n-1} \le \int_{B_r - B_\rho} |\langle \frac{x}{|x|^n}, Df \rangle|.$$

In general all the inequalities in this chapter have different forms, which are easily derived from those stated, and coinciding for almost every value of ρ, r etc.

5.5   Lemma: *Suppose* $f \in BV(B_R)$ *and that* $\rho < R$. *If* $\{\rho_j\}$ *is a sequence such that* $\rho_j \le \rho$ *and* $\rho_j \to \rho$ *then*

(5.3)
$$\begin{cases} \lim_{j \to \infty} \psi(f, \rho_j) = \psi(f, \rho). \\[2mm] \lim_{j \to \infty} v(f, \rho_j) = v(f, \rho), \end{cases}$$

Proof: Given $\varepsilon > 0$, by the definition of $v(f, \rho)$ we can choose a function $g \in BV(B_\rho)$ such that $\mathrm{spt}(g - f) \subset B_\rho$ and

$$\int_{B_\rho} |Dg| \le v(f, \rho) + \varepsilon.$$

For j large enough we have $\mathrm{spt}(g - f) \subset B_{\rho_j}$ and hence

$$\int_{B_\rho} |Dg| \ge \int_{B_{\rho_j}} |Dg| \ge v(f, \rho_j).$$

Since $\varepsilon > 0$ is arbitrary,

$$v(f, \rho) \ge \limsup_{j \to \infty} v(f, \rho_j).$$

On the other hand, for $j \in \mathbb{N}$, we can choose $g_i \in BV(B_\rho)$ such that $\mathrm{spt}(g_j - f) \subset B_{\rho_j}$ and

$$v(f, \rho_j) + \frac{1}{j} \ge \int_{B_{\rho_j}} |Dg_j|.$$

Hence

$$\int\limits_{B_{\rho_j}} |Dg_j| = \int\limits_{B_\rho} |Dg_j| - \int\limits_{B_\rho - B_{\rho_j}} |Df| \geq v(f, \rho) - \int\limits_{B_\rho - B_{\rho_j}} |Df|$$

and therefore

$$\liminf_{j \to \infty} v(f, \rho_j) \geq v(f, \rho). \qquad \square$$

5.6 Lemma: *Suppose* $f, g \in BV(B_R)$ *and* $\rho < R$. *Then*

(5.4) $|v(f, \rho) - v(g, \rho)| \leq \int\limits_{\partial B_\rho} |f^- - g^-| dH_{n-1}.$

Proof: As (5.4) is symmetric in f, g it is sufficient to show that

$$v(g, \rho) - v(f, \rho) \leq \int\limits_{\partial B_\rho} |f^- - g^-| dH_{n-1}.$$

Now given $\varepsilon > 0$, we can choose $\varphi \in BV(B_R)$ such that $\mathrm{spt}(\varphi - f) \subseteq B_\rho$ and

$$\int\limits_{B_\rho} |D\varphi| \leq v(f, \rho) + \varepsilon.$$

Let $\{\rho_j\}$ be a sequence such that $\rho_j \leq \rho$, $\rho_j \to \rho$,

$$\int\limits_{\partial B_{\rho_j}} |Df| = \int\limits_{\partial B_{\rho_j}} |Dg| = 0$$

and $\mathrm{spt}(f - \varphi) \subseteq B_{\rho_j}$. For every j, define

$$g_j = \begin{cases} \varphi \text{ in } B_{\rho_j} \\ g \text{ in } B_R - \bar{B}_{\rho_j}. \end{cases}$$

Then, by Proposition 2.8, $g_j \in BV(B_R)$ and

$$v(g, \rho) \leq \int\limits_{B_\rho} |Dg_j| = \int\limits_{B_{\rho_j}} |D\varphi| + \int\limits_{B_\rho - B_{\rho_j}} |Dg| + \int\limits_{\partial B_{\rho_j}} |f - g| dH_{n-1} \leq$$

$$\leq \int\limits_{B_\rho} |D\varphi| + \int\limits_{B_\rho - B_{\rho_j}} |Dg| + \int\limits_{\partial B_{\rho_j}} |f - g| dH_{n-1} \leq$$

$$\le v(f,\,\rho) + \varepsilon + \int\limits_{B_\rho - B_{\rho_j}} |Dg| + \int\limits_{\partial B_{\rho_j}} |f - g| \, dH_{n-1}.$$

Now as $\varepsilon > 0$ is arbitrary the lemma follows by allowing $j \to \infty$. $\qquad\square$

5.7 **Remark:** In particular if $\psi(f,\,R) = 0$, we have

$$v(f,\,\rho) = \int\limits_{B_\rho} |Df|$$

and hence, for every $g \in BV(B_\rho)$,

$$(5.5) \quad \int\limits_{B_\rho} |Df| \le \int\limits_{B_\rho} |Dg| + \int\limits_{\partial B_\rho} |f^- - g^-| \, dH_{n-1}.$$

5.8 **Lemma:** *Suppose* $f \in BV(B_R)$ *and* $0 < \rho < r < R$. *Then*

$$(5.6) \quad \{ \int\limits_{B_r - B_\rho} |\langle \frac{x}{|x|^n},\, Df \rangle|\}^2 \le 2\{ \int\limits_{B_r - B_\rho} |x|^{1-n} |Df| \}$$

$$\{ r^{1-n} \int\limits_{B_r} |Df| - \rho^{1-n} \int\limits_{B_\rho} |Df| + (n-1) \int\limits_\rho^r t^{-n} \psi(f,\,t) \, dt \}.$$

Proof: Suppose first that $f \in C^1(B_R)$ and then, for $0 < t < R$, define

$$f_t(x) = \begin{cases} f(x) & t < |x| < R \\ f\left(t\dfrac{x}{|x|} \right) & |x| < t. \end{cases}$$

Then we have

$$\int\limits_{B_t} |Df_t| \, dx = \frac{t}{n-1} \int\limits_{\partial B_t} |Df| \left\{ 1 - \frac{\langle x,\, Df \rangle^2}{|x|^2 |Df|^2} \right\}^{1/2} dH_{n-1}$$

and hence

$$v(f,\,t) = \int\limits_{B_t} |Df| - \psi(f,\,t) \le \int\limits_{B_t} |Df_t| \le$$

$$\le \frac{t}{n-1} \int\limits_{\partial B_t} |Df| \, dH_{n-1} - \frac{t}{2(n-1)} \int\limits_{\partial B_t} \frac{\langle x,\, Df \rangle^2}{|x|^2 |Df|} \, dH_{n-1}$$

and

$$\frac{1}{2}t^{1-n}\int_{\partial B_t}\frac{\langle x, Df\rangle^2}{|x|^2|Df|}dH_{n-1} \leqq t^{1-n}\int_{\partial B_t}|Df|dH_{n-1} -$$

$$- (n-1)t^{-n}\int_{B_t}|Df|dx + (n-1)t^{-n}\psi(f, t) =$$

$$= \frac{d}{dt}(t^{1-n}\int_{B_t}|Df|dx) + (n-1)t^{-n}\psi(f, t).$$

Now integrating with respect to t between ρ and r, we have

$$\frac{1}{2}\int_{B_r - B_\rho}\frac{\langle x, Df\rangle^2}{|x|^{n+1}|Df|}dx \leqq r^{1-n}\int_{B_R}|Df|dx - \rho^{1-n}\int_{B_\rho}|Df|dx +$$

$$+ (n-1)\int_\rho^r t^{-n}\psi(f, t)dt.$$

On the other hand, from the Schwartz inequality we have

$$\left\{\int_{B_r - B_\rho}|\langle\frac{x}{|x|^n}, Df\rangle|dx\right\}^2 \leqq \int_{B_r - B_\rho}|x|^{1-n}|Df|dx \int_{B_r - B_\rho}\frac{\langle x, Df\rangle^2}{|x|^{n+1}|Df|}dx$$

and so (5.6) holds for $f \in C^1(B_R)$.

Now suppose that $f \in BV(B_R)$; then by Remarks 2.12 and 2.13 we can approximate f by C^1 functions f_k such that for almost all t we have

$$\int_{B_t}|Df_k|dx \to \int_{B_t}|Df|$$

and

$$\int_{\partial B_t}|f - f_k|dH_{n-1} \to 0.$$

If we write (5.6) for f_k and observe that, by Lemma 5.6, $\psi(f_k, t) \to \psi(f, t)$, we see that (5.6) holds for $f \in BV(B_R)$ and almost all ρ, r. Finally we obtain (5.6) for every ρ and r by approximating with increasing sequences $\{\rho_j\} \to \rho$ and $\{r_j\} \to r$ for which (5.6) holds.   □

5.9   Remark: By approximating at the final step with sequences decreasing to r and ρ we obtain (5.6) with $\bar{B}_r$, $\bar{B}_\rho$ instead of B_r and B_ρ.

5.10   Remark: From (5.6) it follows that, for every $\rho < r$,

$$(5.7) \quad \rho^{1-n} \int_{\dot{B}_\rho} |Df| \leq r^{1-n} \int_{\dot{B}_r} |Df| + (n-1) \int_\rho^r t^{-n} \psi(f, t)dt$$

and in particular, if $\psi(f, r) = 0$, then

$$(5.8) \quad \rho^{1-n} \int_{\dot{B}_\rho} |Df| \leq r^{1-n} \int_{\dot{B}_r} |Df|.$$

Hence $\rho^{1-n} \int_{\dot{B}_\rho} |Df|$ is an increasing function of ρ.

5.11   Lemma: *Suppose* $f \in BV(B_R)$ *and* $0 < \rho < r < R$; *then*

$$(5.9) \quad \int_{B_r - B_\rho} |x|^{1-n} |Df| \leq \left\{ 1 + (n-1)\log\frac{r}{\rho} \right\} r^{1-n} \int_{\dot{B}_r} |Df| +$$

$$+ (n-1)^2 \int_\rho^r s^{-n} \log\frac{s}{\rho} \psi(f, s)ds.$$

Proof: As usual, it is sufficient to prove (5.9) for $f \in C^1(B_R)$, in which case we have

$$\int_{B_r - B_\rho} |x|^{1-n} |Df|dx = \int_\rho^r t^{1-n} \Big(\int_{\partial B_t} |Df|dH_{n-1} \Big)dt = \int_\rho^r t^{1-n} \eta'(t)dt$$

where

$$\eta(t) = \int_{\dot{B}_t} |Df|dx.$$

Integrating by parts,

$$\int_\rho^r t^{1-n} \eta'(t)dt \leq r^{1-n} \int_{\dot{B}_r} |Df|dx + (n-1) \int_\rho^r t^{-n} \Big(\int_{\dot{B}_t} |Df|dx \Big)dt.$$

We now estimate this last integral to obtain the result. Indeed,

$$t^{-n}\int_{\dot B_t}|Df|\,dx \leq t^{-1}\{r^{1-n}\int_{\dot B_r}|Df|\,dx + (n-1)\int_t^r s^{-n}\psi(f,s)ds\}$$

(which follows from (5.7)) and hence

$$\int_\rho^r t^{-n}dt\int_{\dot B_t}|Df|\,dx \leq r^{1-n}\log\frac{r}{\rho}\int_{\dot B_r}|Df|\,dx + (n-1)\int_\rho^r\frac{dt}{t}\int_t^r s^{-n}\psi(f,s)ds =$$

$$= r^{1-n}\log\frac{r}{\rho}\int_{\dot B_r}|Df|\,dx + (n-1)\int_\rho^r s^{-n}\log\frac{s}{\rho}\psi(f,s)ds$$

from which (5.9) follows at once.   □

5.12   Proposition: *Suppose* $f\in BV(B_R)$ *and* $0 < \rho < r < R;$ *then*

(5.10) $|r^{1-n}\int_{\dot B_r}Df - \rho^{1-n}\int_{\dot B_\rho}Df|^2 \leq \{2r^{1-n}\left(1+(n-1)\log\frac{r}{\rho}\right)\int_{\dot B_r}|Df| +$

$$+ 2(n-1)^2\int_\rho^r s^{-n}\log\frac{s}{\rho}\psi(f,s)ds\}$$

$$\{r^{1-n}\int_{\dot B_r}|Df| - \rho^{1-n}\int_{\dot B_\rho}|Df| + (n-1)\int_\rho^r s^{-n}\psi(f,s)ds\}.$$

Proof: Inequality (5.10) follows easily from (5.1), (5.6) and (5.9) by noting that, from Remark 2.13,

$$\int_{\dot B_t}Df = \int_{\partial B_t}f^-(x)\frac{x}{|x|}dH_{n-1} = t^{n-1}\int_{\partial B_1}f^-(tx)x\,dH_{n-1}$$

and hence

$$|r^{1-n}\int_{\dot B_r}Df - \rho^{1-n}\int_{\dot B_\rho}|Df| \leq \int_{\partial B_1}|f^-(rx) - f^-(\rho x)|dH_{n-1}$$ □

5.13   Remark: A particular case of interest is when $f = \varphi_E$ and E is a set of minimizing boundary in B_R, that is

$$\psi(E, R) = 0.$$

In this case we have

$$(5.11) \quad \{\int_{\partial B_1} |\varphi_E^-(\rho x) - \varphi_E^-(rx)| dH_{n-1}\}^2 \leq \{\int_{B_r - B_\rho} |\langle \frac{x}{|x|^n}, D\varphi_E\rangle|\}^2$$

$$\leq 2 \int_{B_r - B_\rho} |x|^{1-n} |D\varphi_E| \{r^{1-n} \int_{B_r} |D\varphi_E| - \rho^{1-n} \int_{B_\rho} |D\varphi_E|\}$$

and hence, for every $\rho < r < R$,

$$(5.12) \quad \rho^{1-n} \int_{B_\rho} |D\varphi_E| \leq r^{1-n} \int_{B_r} |D\varphi_E|.$$

Moreover

$$(5.13) \quad |r^{1-n} \int_{B_r} D\varphi_E - \rho^{1-n} \int_{B_\rho} D\varphi_E|^2 \leq 2\left(1 + (n-1)\log\frac{r}{\rho}\right) r^{1-n} \int_{B_r} |D\varphi_E|$$

$$\{r^{1-n} \int_{B_r} |D\varphi_E| - \rho^{1-n} \int_{B_\rho} |D\varphi_E|\}.$$

Now choose $0 < s < r \leq R$ and consider the sets $E - \bar{B}_s$ and $E \cup \bar{B}_s$. By the definition of $\psi(E, R)$ we have

$$P(E - \bar{B}_s, B_r) = \int_{B_r} |D\varphi_{E - \bar{B}_s}| \geq \int_{B_r} |D\varphi_E|$$

and

$$P(E \cup \bar{B}_s, B_r) = \int_{B_r} |D\varphi_{E \cup \bar{B}_s}| \geq \int_{B_r} |D\varphi_E|.$$

As in (2.17) and (2.18) of Remark 2.14, we obtain

$$P(E - \bar{B}_s, B_r) = P(E, B_r - \bar{B}_s) + H_{n-1}(\partial B_s \cap E)$$

and

$$P(E \cup \bar{B}_s, B_r) = P(E, B_r - \bar{B}_s) + H_{n-1}(\partial B_s \cap (\mathbb{R}^n - E))$$

for almost all $s < r$. Thus

$$P(E, B_r) \leqq P(E, B_r - \bar{B}_s) + \min\{H_{n-1}(\partial B_s \cap E), H_{n-1}(\partial B_s \cap (\mathbb{R}^n - E))\} \leqq$$

$$\leqq P(E, B_r - \bar{B}_s) + \frac{1}{2}ns^{n-1}\omega_n.$$

for almost all s. Now choosing a sequence $\{s_j\}$ such that the above inequality holds and $s_j \to r$ as $j \to \infty$, we obtain

$$(5.14) \quad \int_{B_r} |D\varphi_E| = v(E, r) \leqq \frac{1}{2}n\omega_n r^{n-1}.$$

Using this in (5.13) gives

$$(5.15) \quad |r^{1-n} \int_{B_R} D\varphi_E - \rho^{1-n} \int_{B_\rho} D\varphi_E|^2 \leqq n\omega_n\left(1 + (n-1)\log\frac{r}{\rho}\right)$$

$$\{r^{1-n} \int_{B_r} |D\varphi_E| - \rho^{1-n} \int_{B_\rho} |D\varphi_E|\}.$$

Finally, letting $\rho \to 0$ in (5.12) and recalling (3.17) we have

$$(5.16) \quad \omega_{n-1}r^{n-1} \leqq \int_{B(x,r)} |D\varphi_E|$$

for every $x \in \partial^* E$ and hence, as $\partial E = \overline{\partial^* E}$, (5.16) holds for every x in ∂E by approximation.

A similar inequality holds for the volume of $E \cap B(x, r)$.

5.14 Proposition: *Suppose* $\psi(E, \Omega) = 0$, *and let* $x_0 \in E$. *Then for every* $r < \mathrm{dist}(x_0, \partial\Omega)$ *we have:*

$$(5.17) \quad |E \cap B(x_0, r)| \geqq \frac{r^n}{2nc_1(n)}$$

where $c_1(n)$ *is the isoperimetric constant of Corollary 1.29.*

Proof: Let $\rho < \mathrm{dist}(x_0, \partial\Omega)$. We have

$$\int_{\Omega} |D\varphi_E| \leqq \int_{\Omega} |D\varphi_{E-B_\rho}|$$

and hence

$$\int_{B_\rho} |D\varphi_E| \leqq \int_{\partial B_\rho} \varphi_E dH_{n-1}.$$

On the other hand, for almost every ρ there holds:

$$\int |D\varphi_{E \cap B_\rho}| = \int_{B_\rho} |D\varphi_E| + \int_{\partial B_\rho} \varphi_E dH_{n-1}$$

and therefore, setting $E_\rho = E \cap B_\rho$:

$$\int |D\varphi_{E_\rho}| \leqq 2 \int_{\partial B_\rho} \varphi_E dH_{n-1} = 2\frac{d}{d\rho}|E_\rho|.$$

From the isoperimetric inequality (1.10) we get

$$\frac{d}{d\rho}|E_\rho| \geqq \frac{1}{2c_1(n)}|E_\rho|^{\frac{n-1}{n}}$$

and integrating:

$$|E_r| \geqq r^n/2nc_1(n). \qquad \qquad \square$$

5.15   Remark: Since $\Omega - E$ also minimizes perimeter in Ω, if $x_0 \in \partial E$ we have

$$(5.18) \quad a|B_r| \leqq |E_r| \leqq (1-a)|B_r|$$

with $a^{-1} = 2n\omega_n c_1(n)$.

6. Approximation of Minimal Sets (I)

The most important step in the regularity theory for minimal sets is the following lemma of De Giorgi ([DG5], [MM3]).

De Giorgi Lemma: *For every $n \geq 2$ and for every α, $0 < \alpha < 1$, there exists a positive constant $\sigma = \sigma(n, \alpha)$ such that, if E is a Caccioppoli set in $\mathbb{R}^n$ and for some $\rho > 0$*

$$\psi(E, \rho) = 0,$$

$$\int_{B_\rho} |D\varphi_E| - |\int_{B_\rho} D\varphi_E| < \sigma(n, \alpha)\rho^{n-1},$$

then

$$\int_{B_{\alpha\rho}} |D\varphi_E| - |\int_{B_{\alpha\rho}} D\varphi_E| < \alpha^n \{ \int_{B_\rho} |D\varphi_E| - |\int_{B_\rho} D\varphi_E| \}.$$

This will in fact appear in chapter 8 as Theorem 8.1.

It can be seen that the term

$$\Lambda(E, \rho) = \rho^{1-n} \{ \int_{B_\rho} |D\varphi_E| - |\int_{B_\rho} D\varphi_E| \}$$

is of fundamental importance in the regularity theory, particularly in the way it varies as ρ varies. Indeed some writers give it a special name, calling it the *Excess*. By the properties of the reduced boundary we can also write

$$\Lambda(E, \rho) = \rho^{1-n} \{ H_{n-1}(B_\rho \cap \partial^* E) - | \int_{B_\rho \cap \partial^* E} v(x) dH_{n-1} | \},$$

that is $\Lambda(E, \rho)$ is a measure of how much the direction of $v(x)$ changes in $B_\rho \cap \partial^* E$ (we already know that $|v(x)| = 1$ on $B_\rho \cap \partial^* E$). If $\Lambda(E, \rho)$ is small, then $v(x)$ must remain approximately in a constant direction and we can expect to apply Theorem 4.8 or some similar result.

To prove the lemma, we notice that if $E \cap B_\rho$ can be written as

$$E \cap B_\rho = \{(x, t): x \in A, \, t > f(x)\} \cap B_\rho$$

where $f \in C^1(A)$ and $A \subseteq \mathbb{R}^{n-1}$, then

$$\int_{\dot{B}_\rho} |D\varphi_E| = \int_A \sqrt{1 + |Df|^2} \, dx.$$

Hence we are considering the problem of minimizing the integral on the right hand side among functions in $C^1(A)$.

Now if $|Df|$ is small, that is ∂E is nearly flat in B_ρ, then $\sqrt{1 + |Df|^2}$ is approximately equal to $1 + \frac{1}{2}|Df|^2$ and so f must almost minimize the integral

$$I(f) = \int_A |Df|^2 dx,$$

that is f must be nearly harmonic. However estimates like those of the De Giorgi Lemma are available for harmonic functions (see Lemma 6.1 below) and so the idea of the proof is as follows. Assume ∂E is C^1 and nearly flat. Then by approximating with harmonic functions we prove an inequality as in the De Giorgi Lemma. If ∂E is not C^1 then we approximate by surfaces which are C^1 but may no longer be minimal, only close to minimal.

It can be seen that it is important to be able to obtain estimates for harmonic functions which approximate sequences of surfaces tending to a minimum in some sense. We shall do this in the present chapter. It is also important to be able to obtain C^1 hypersurfaces approximating any given minimal surfaces and then obtain estimates as they change. This will be done in Chapter 7.

We now prove the analogue of the De Giorgi Lemma for harmonic functions.

6.1 Lemma: *Suppose $\mathscr{B}_\rho \subseteq \mathbb{R}^m$ and $u \in C^1(\mathscr{B}_\rho)$ is harmonic in $\mathscr{B}_\rho$ (that is*

$$\sum_{i=1}^m \frac{\partial^2}{\partial x_i^2} u(x) = 0 \text{ for } x \in \mathscr{B}_\rho) \text{ and let}$$

$$q = \frac{1}{|\mathscr{B}_\rho|} \int_{\mathscr{B}_\rho} Du \, dx.$$

Then for every α, $0 < \alpha < 1$,

(6.1) $\displaystyle \int_{\mathscr{B}_{\alpha\rho}} (|Du|^2 - |q|^2) dx \leq \alpha^{m+2} \int_{\mathscr{B}_\rho} (|Du|^2 - |q|^2) dx.$

Proof: The harmonic function u may be written as a sum (possibly infinite) of homogeneous orthogonal harmonic polynomials (see [FJ], [AGMMP]). Thus

(i) $u = \sum\limits_{i=0}^{\infty} V_i$

where V_i is a harmonic polynomial of degree i,

(ii) $\int\limits_{\mathscr{B}_{\alpha\rho}} <DV_j, DV_k> dx = \int\limits_{\mathscr{B}_{\rho}} <DV_j, DV_k> dx = 0$

if $k \neq j$, and

(iii) $\int\limits_{\mathscr{B}_{\alpha\rho}} DV_j \, dx = \int\limits_{\mathscr{B}_{\rho}} DV_j \, dx = 0$

if $j \geq 2$. Therefore $q = DV_1$ and

$$\int_{\mathscr{B}_{\rho}} (|Du|^2 - |q|^2) dx = \sum_{j=2}^{\infty} \int_{\mathscr{B}_{\rho}} |DV_j|^2 dx,$$

$$\int_{\mathscr{B}_{\alpha\rho}} (|Du|^2 - |q|^2) dx = \sum_{j=2}^{\infty} \int_{\mathscr{B}_{\alpha\rho}} |DV_j|^2 dx.$$

(6.1)   now follows from the homogeneity of V_j.   □

We now show that, if we have a sequence of C^1 functions whose gradients tend to zero and which do not differ too much from the corresponding harmonic functions (in the sense that the areas of the defined surfaces are close), then we may prove an inequality like (6.1).

6.2 Lemma: *Suppose* $\{\omega_j\}$ *is a sequence in* $C^1(\bar{\mathscr{B}}_{\rho})$, *and* $\{\beta_j\}$ *is a sequence of positive numbers. For* $j \in \mathbb{N}$ *let* u_j *be the harmonic function in* $\mathscr{B}_{\rho}$ *such that* $u_j = \omega_j$ *on* $\partial\mathscr{B}_{\rho}$ *and, for any function* $f \in C(\mathscr{B}_{\rho})$ *and any* $r \leq \rho$, *let*

$$\{f\}_r = \frac{1}{|\mathscr{B}_r|} \int_{\mathscr{B}_r} f dx;$$

that is $\{f\}_r$ *is the mean value of* f *on* $\mathscr{B}_r$.

Suppose that

(6.2) $\lim_{j \to \infty} \sup_{\mathscr{B}_\rho} |D\omega_j| = 0,$

(6.3) $\int_{\mathscr{B}_\rho} \{\sqrt{1 + |D\omega_j|^2} - \sqrt{1 + |\{D\omega_j\}_\rho|^2}\} dx \leqq \beta_j,$

(6.4) $\lim_{j \to \infty} \sup \beta_j^{-1} \int_{\mathscr{B}_\rho} \{\sqrt{1 + |D\omega_j|^2} - \sqrt{1 + |Du_j|^2}\} dx = 0.$

Then for every α, $0 < \alpha < 1$,

(6.5) $\lim_{j \to \infty} \sup \beta_j^{-1} \int_{\mathscr{B}_{\alpha\rho}} \{\sqrt{1 + |D\omega_j|^2} - \sqrt{1 + |\{D\omega_j\}_{\alpha\rho}|^2}\} dx \leqq \alpha^{m+2}.$

Proof: From the Taylor expansion of $\sqrt{1 + x}$ about B^2 we have

$$\sqrt{1 + A^2} - \sqrt{1 + B^2} - \frac{(A^2 - B^2)}{2\sqrt{1 + B^2}} = -\frac{(A^2 - B^2)^2}{8(1 + \xi^2)^{3/2}}$$

for some ξ between A and B. Hence

(6.6) $\sqrt{1 + A^2} - \sqrt{1 + B^2} \leqq \frac{(A^2 - B^2)}{2\sqrt{1 + B^2}}$

and, if $B^2 < 1$, it easily follows that

(6.7) $\sqrt{1 + A^2} - \sqrt{1 + B^2} - \frac{(A^2 - B^2)}{2\sqrt{1 + B^2}} \geqq -\frac{(A^2 - B^2)^2}{2\sqrt{1 + B^2}}.$

From (6.6) we obtain

(6.8) $\int_{\mathscr{B}_{\alpha\rho}} \{\sqrt{1 + |D\omega_j|^2} - \sqrt{1 + |\{D\omega_j\}_{\alpha\rho}|^2}\} dx \leqq$

$$\leqq \frac{1}{2\sqrt{1 + |\{D\omega_j\}_{\alpha\rho}|^2}} \int_{\mathscr{B}_{\alpha\rho}} [|D\omega_j|^2 - |\{D\omega_j\}_{\alpha\rho}|^2] dx.$$

Now, by definition of $\{D\omega_j\}_{\alpha\rho}$,

$$\int_{\mathscr{B}_{\alpha\rho}} [|D\omega_j|^2 - |\{D\omega_j\}_{\alpha\rho}|^2] dx = \int_{\mathscr{B}_{\alpha\rho}} |D\omega_j - \{D\omega_j\}_{\alpha\rho}|^2 dx \leqq$$

$$\leqq \int_{\mathscr{B}_{\alpha\rho}} |D\omega_j - \{D\omega_j\}_\rho|^2 dx,$$

and so

(6.9) $\displaystyle \limsup_{j\to\infty} \beta_j^{-1} \int_{\mathscr{B}_{\alpha\rho}} \{\sqrt{1 + |D\omega_j|^2} - \sqrt{1 + |\{D\omega_j\}_{\alpha\rho}|^2}\} dx \leqslant$

$$\leqq \frac{1}{2} \limsup_{j\to\infty} \beta_j^{-1} \int_{\mathscr{B}_{\alpha\rho}} |D\omega_j - \{D\omega_j\}_\rho|^2 dx.$$

We must now estimate the right hand side of (6.9). If $A, B, C \in \mathbb{R}^m$, then by the triangle inequality and Cauchy's inequality

$$|A - B|^2 \leqq (1 + \tfrac{1}{\varepsilon})|A - C|^2 + (1 + \varepsilon)|B - C|^2$$

for every $\varepsilon > 0$. Therefore, for $\varepsilon > 0$,

(6.10) $\displaystyle \int_{\mathscr{B}_{\alpha\rho}} |D\omega_j - \{D\omega_j\}_\rho|^2 dx \leqq (1 + \tfrac{1}{\varepsilon}) \int_{\mathscr{B}_{\alpha\rho}} |D\omega_j - Du_j|^2 dx$

$$+ (1 + \varepsilon) \int_{\mathscr{B}_{\alpha\rho}} |Du_j - \{D\omega_j\}_\rho|^2 dx.$$

By the properties of harmonic functions

$$\{Du_j\}_{\alpha\rho} = \{Du_j\}_\rho = \{D\omega_j\}_\rho$$

and hence, from Lemma 6.1,

$$\int_{\mathscr{B}_{\alpha\rho}} |Du_j - \{D\omega_j\}_\rho|^2 dx \leqq \alpha^{m+2} \int_{\mathscr{B}_\rho} |Du_j - \{D\omega_j\}_\rho|^2 dx.$$

Moreover

$$\int_{\mathscr{B}_\rho} |Du_j - \{D\omega_j\}_\rho|^2 dx \leqq (1 + \varepsilon) \int_{\mathscr{B}_\rho} |D\omega_j - \{D\omega_j\}_\rho|^2 dx +$$

$$+ (1 + \tfrac{1}{\varepsilon}) \int_{\mathscr{B}_\rho} |D\omega_j - Du_j|^2 dx$$

and so from (6.10) we have

(6.11) $\int_{\mathscr{B}_{\alpha\rho}} |D\omega_j - \{D\omega_j\}_\rho|^2 dx \leqq \alpha^{m+2}(1+\varepsilon)^2 \int_{\mathscr{B}_\rho} |D\omega_j - \{D\omega_j\}_\rho|^2 dx +$

$$+ Q \int_{\mathscr{B}_\rho} |D\omega_j - Du_j|^2 dx,$$

where Q depends only on ε, α and m.

We now use (6.2), (6.3) and (6.4) to estimate the second term on the right hand side of (6.11). We have from (6.7)

$$\int_{\mathscr{B}_\rho} \left(\sqrt{1 + |D\omega_j|^2} - \sqrt{1 + |\{D\omega_j\}_\rho|^2} \right) dx \geqslant$$

$$\geqq \frac{1}{2\sqrt{1 + |\{D\omega_j\}_\rho|^2}} \left\{ \int_{\mathscr{B}_\rho} (|D\omega_j|^2 - |\{D\omega_j\}_\rho|^2) dx - \right.$$

$$\left. - \int_{\mathscr{B}_\rho} (|D\omega_j|^2 - |\{D\omega_j\}_\rho|^2)^2 dx \right\}.$$

Now

$$(|D\omega_j|^2 - |\{D\omega_j\}_\rho|^2)^2 \leqq |D\omega_j - \{D\omega_j\}_\rho|^2 \left(\sup_{\mathscr{B}_\rho} |D\omega_j| + |\{D\omega_j\}|_\rho \right)^2 =$$

$$= m_j |D\omega_j - \{D\omega_j\}_\rho|^2$$

and hence

(6.12) $\int_{\mathscr{B}_\rho} \left(\sqrt{1 + |D\omega_j|^2} - \sqrt{1 + |\{D\omega_j\}_\rho|^2} \right) dx \geqslant$

$$\geqq \frac{1 - m_j}{2\sqrt{1 + |\{D\omega_j\}_\rho|^2}} \int_{\mathscr{B}_\rho} (|D\omega_j|^2 - |\{D\omega_j\}_\rho|^2) dx.$$

From (6.12), (6.2) and (6.3) we deduce that

(6.13) $\displaystyle\limsup_{j \to \infty} \beta_j^{-1} \int_{\mathscr{B}_\rho} (|D\omega_j|^2 - |\{D\omega_j\}_\rho|^2) dx \leqq 2.$

On the other hand from (6.6)

$$\int_{\mathscr{B}_\rho} (|\{D\omega_j\}_\rho|^2 - |Du_j|^2)dx \leqslant$$

$$\leqq 2\sqrt{1 + |\{D\omega_j\}_\rho|^2} \int_{\mathscr{B}_\rho} (\sqrt{1 + |\{D\omega_j\}_\rho|^2} - \sqrt{1 + |Du_j|^2})dx.$$

Comparing this with (6.12) and observing, as u is harmonic, that

$$\int_{\mathscr{B}_\rho} |D\omega_j - Du_j|^2 dx = \int_{\mathscr{B}_\rho} (|D\omega_j|^2 - |Du_j|^2)dx,$$

we conclude

$$\int_{\mathscr{B}_\rho} |D\omega_j - Du_j|^2 dx \leqq 2\sqrt{1 + |\{D\omega_j\}_\rho|^2} \left[\int_{\mathscr{B}_\rho} (\sqrt{1 + |D\omega_j|^2} - \right.$$

$$\left. - \sqrt{1 + |Du_j|^2})dx + \frac{m_j}{1 - m_j} \int_{\mathscr{B}_\rho} (\sqrt{1 + |D\omega_j|^2} - \sqrt{1 + |\{D\omega_j\}_\rho|^2})dx \right].$$

Hence

$$\lim_{j \to \infty} \beta_j^{-1} \int_{\mathscr{B}_\rho} |D\omega_j - Du_j|^2 dx = 0.$$

This fact, together with (6.13), (6.11) and (6.9), gives

$$\limsup_{j \to \infty} \beta_j^{-1} \int_{\mathscr{B}_{\alpha\rho}} [\sqrt{1 + |D\omega_j|^2} - \sqrt{1 + |\{D\omega_j\}_{\alpha\rho}|^2}]dx \leqq (1 + \varepsilon)^2 \alpha^{m+2}$$

and, allowing $\varepsilon \to 0$, (6.5) holds. $\square$

The next lemma is similar to the last one except that now, rather than considering functions satisfying (6.4), we shall look at sets determined by smooth functions and replace (6.4) with a condition which says that the sets tend to a minimum.

If $\mathscr{B}_\rho \subseteq \mathbb{R}^m$ and $\omega \in C^1(\mathscr{B}_\rho)$, we define

$$W = \{(x, t): x \in \mathscr{B}_\rho, t < \omega(x)\},$$

$$Q = \{(x, t): x \in \mathscr{B}_\rho, \min \omega - 1 < t < \max \omega + 1\}.$$

6.3 **Lemma:** *Suppose* $\{\omega_j\}$ *is a sequence in* $C^1(\bar{\mathscr{B}}_\rho)$ *and* $\{\beta_j\}$ *a sequence of positive numbers such that*

$$(6.14) \quad \int_{\mathscr{B}_\rho} (\sqrt{1 + |D\omega_j|^2} - \sqrt{1 + |\{D\omega_j\}_\rho|^2})dx \leq \beta_j,$$

$$(6.15) \quad \limsup_{j\to\infty \mathscr{B}_\rho} |D\omega_j| = 0,$$

$$(6.16) \quad \lim_{j\to\infty} \beta_j^{-1} \, \psi(W_j, Q_j) = 0.$$

Then for every α, $0 < \alpha < 1$,

$$(6.17) \quad \limsup_{j\to\infty} \beta_j^{-1} \int_{\mathscr{B}_{\alpha\rho}} (\sqrt{1 + |D\omega_j|^2} - \sqrt{1 + |\{D\omega_j\}_{\alpha\rho}|^2})dx \leq \alpha^{m+2}.$$

Proof: Denote by u_j the harmonic function in $\mathscr{B}_\rho$ which is equal to ω_j on $\partial\mathscr{B}_\rho$. Then

$$\int_{\mathscr{B}_\rho} (\sqrt{1 + |D\omega_j|^2} - \sqrt{1 + |Du_j|^2})dx \leq \psi(W_j, Q_j)$$

and the conclusion follows from Lemma 6.2.   □

We now prove a similar lemma, but it now concerns Caccioppoli sets with the appropriate conditions and uses the previous lemma with $m = n - 1$.

6.4 **Lemma:** *Suppose* $\{L_j\}$ *is a sequence of Caccioppoli sets in* $\mathbb{R}^n$, $\{\beta_j\}$ *is a sequence of positive numbers and* $\rho > 0$ *is such that*

$$(6.18) \quad \int_{B_\rho} |D\varphi_{L_j}| - |\int_{B_\rho} D\varphi_{L_j}| \leq \beta_j,$$

$$(6.19) \quad \partial L_j \cap B_\rho \text{ is a } C^1\text{-hypersurface and}$$

$$\liminf_{j\to\infty \ \partial L_j \cap B_\rho} v_n^j(x) = 1$$

($v^j(x)$ is the normal to L_j at the point x),

(6.20) $\displaystyle\lim_{j\to\infty} \beta_j^{-1}\,\psi(L_j,\,\rho) = 0.$

Then for every α, $0 < \alpha < 1$,

(6.21) $\displaystyle\limsup_{j\to\infty}\beta_j^{-1}\left\{\int_{B_{\alpha\rho}}|D\varphi_{L_j}| - |\int_{B_{\alpha\rho}}D\varphi_{L_j}|\right\} \leq \alpha^{n+1}$

Proof: Suppose there exists a sequence $\{L_j\}$ and a number α for which (6.18), (6.19) and (6.20) hold but

(6.22) $\displaystyle\lim_{j\to\infty}\beta_j^{-1}\left\{\int_{B_{\alpha\rho}}|D\varphi_{L_j}| - |\int_{B_{\alpha\rho}}D\varphi_{L_j}|\right\} > \alpha^{n+1}.$

We may suppose that $v_n^j(x) \geq \frac{1}{2}$ for every $x \in \partial L_j \cap B$ and every j. By Theorems 4.8 and 4.11 we may conclude that there exist open sets $A_j \subseteq \mathbb{R}^{n-1}$ and C^1-functions $\omega_j : A_j \to \mathbb{R}$ such that

$$\partial L_j \cap B_\rho = \{(y,\,t) \in \mathbb{R}^n : y \in A_j,\; t = \omega_j(y)\}$$

and by (4.14) and (6.19)

(6.23) $\displaystyle\limsup_{j\to\infty\, A_j}|D\omega_j| = 0.$

Furthermore, since $\displaystyle\sup_{A_j}|\omega_j| \leq \rho$, we may assume (taking a subsequence if necessary) that

(6.24) $\displaystyle\liminf_{j\to\infty\, A_j}\omega_j = c.$

Then we must have $c^2 < \rho^2$, that is we cannot have $c^2 = \rho^2$. Indeed otherwise from (6.23) we would have, for j sufficiently large,

$$\partial L_j \cap B_{\alpha\rho} = \varnothing$$

which contradicts (6.22). From (6.23) and (6.24) we may conclude that for every $\varepsilon > 0$ there exists a j'_ε such that

$$|\omega_j - c| < \varepsilon \quad \text{for} \quad j > j'_\varepsilon$$

and so, from the definition of ω_j, it follows that there exists a j_ε such that if $\sigma^2 = \rho^2 - c^2$ then

(6.25) $\quad \mathscr{B}_{\sigma - \varepsilon} \subseteq A_j \subseteq \mathscr{B}_{\sigma + \varepsilon} \quad \text{for} \quad j > j_\varepsilon.$

From Example 1.4 we have

$$\int_{(\mathscr{B}_{\sigma - \varepsilon} \times \mathbb{R}) \cap B_\rho} D\varphi_{L_j} = \int_{\partial L_j \cap B_\rho} v^j(x) dH_{n-1} \quad \text{for} \quad j > j_\varepsilon.$$

Hence, introducing a change of variables and taking note of Remark 4.10, for $j > j_\varepsilon$,

$$\int_{(\mathscr{B}_{\sigma - \varepsilon} \times \mathbb{R}) \cap B_\rho} D_i\varphi_{L_j} = \int_{\mathscr{B}_{\sigma - \varepsilon}} D_i\omega_j dy \qquad i = 1, 2, \ldots, n-1,$$

$$\int_{(\mathscr{B}_{\sigma - \varepsilon} \times \mathbb{R}) \cap B_\rho} D_n\varphi_{L_j} = |\mathscr{B}_{\sigma - \varepsilon}|,$$

$$\int_{(\mathscr{B}_{\sigma - \varepsilon} \times \mathbb{R}) \cap B_\rho} |D\varphi_{L_j}| = \int_{\mathscr{B}_{\sigma - \varepsilon}} \sqrt{1 + |D\omega_j|^2} \, dy.$$

Therefore

$$\sqrt{1 + |\{D\omega_j\}_{\sigma - \varepsilon}|^2} = \frac{1}{|\mathscr{B}_{\sigma - \varepsilon}|} \left| \int_{(\mathscr{B}_{\sigma - \varepsilon} \times \mathbb{R}) \cap B_\rho} D\varphi_{L_j} \right|$$

and

$$\int_{\mathscr{B}_{\sigma - \varepsilon}} \left(\sqrt{1 + |D\omega_j|^2} - \sqrt{1 + |\{D\omega_j\}_{\sigma - \varepsilon}|^2} \right) dy \leq \int_{B_\rho} |D\varphi_{L_j}| - \left| \int_{B_\rho} D\varphi_{L_j} \right| \leq \beta_j.$$

Now from Lemma 6.3 it follows that for every γ, $0 < \gamma < 1$,

(6.26) $\quad \displaystyle\limsup_{j \to \infty} \beta_j^{-1} \int_{\mathscr{B}_{\gamma(\sigma - \varepsilon)}} \left(\sqrt{1 + |D\omega_j|^2} - \sqrt{1 + |\{D\omega_j\}_{\gamma(\sigma - \varepsilon)}|^2} \right) dy \leq \gamma^{n+1}.$

Let

$$C_j = \{ y \in A_j : (y, \omega_j(y)) \in B_{\alpha\rho} \};$$

then reasoning as for (6.25) we may show that there exists a j'_ε such that, for $j > j'_\varepsilon$,

$$C_j \subseteq \mathcal{B}_{\alpha(\sigma + \varepsilon)}$$

and hence

$$\int\limits_{B_{\alpha\rho}} |D\varphi_{L_j}| - |\int\limits_{B_{\alpha\rho}} D\varphi_{L_j}| \leq \int\limits_{\mathcal{B}_{\alpha(\sigma + \varepsilon)}} (\sqrt{1 + |D\omega_j|^2} - \sqrt{1 + |\{D\omega_j\}_{\alpha(\sigma + \varepsilon)}|^2})\, dy$$

On the other hand taking $\gamma = \alpha\left(\frac{\sigma + \varepsilon}{\sigma - \varepsilon}\right)$ (for ε sufficiently small) in (6.26), we obtain

$$\limsup_{j \to \infty} \beta_j^{-1} \left\{ \int_{B_{\alpha\rho}} |D\varphi_{L_j}| - |\int_{B_{\alpha\rho}} D\varphi_{L_j}| \right\} \leq \alpha^{n+1} \left(\frac{\sigma + \varepsilon}{\sigma - \varepsilon}\right)^{n+1}$$

for all ε sufficiently small. Allowing $\varepsilon \to 0$, we obtain a contradiction to (6.22). $\square$

7. Approximation of Minimal Sets (II)

Our aim in this chapter is to prove a theorem similar to Lemma 6.4 but valid now for arbitrary Caccioppoli sets rather than just sets with C^1-boundary. In order to prove the theorem we approximate with smooth sets and so need fairly detailed estimates of the approximations. Our first choice for C^1-approximations would be the mollified functions introduced in Chapter 1. However, these functions do not produce the required results and so, rather than taking convolutions with positive symmetric mollifiers as introduced in 1.14, we take convolutions with the functions

$$\eta_\varepsilon(x) = \frac{n+1}{\omega_n} \varepsilon^{-n} \max\left\{\left(1 - \frac{|x|}{\varepsilon}\right), 0\right\} = \frac{n+1}{\omega_n} \varepsilon^{-n}\left(1 - \frac{|x|}{\varepsilon}\right) \vee 0.$$

The function η_1 satisfies the conditions for a positive symmetric mollifer except that it is merely Lipschitz continuous instead of C^∞ and spt $\eta_1 = \bar{B}_1$ instead of spt $\eta \subseteq B_1$. Nevertheless many of the properties derived in 1.14 will hold.

For this chapter, given a function $f \in L^1_{\text{loc}}(\mathbb{R}^n)$ and $\varepsilon > 0$, we denote

$$f_\varepsilon(x) = \int \eta_\varepsilon(x - y)f(y)dy.$$

7.1 Lemma: *Suppose E is a Borel set, $\varepsilon > 0$ and let $\varphi_\varepsilon = (\varphi_E)_\varepsilon$. Then φ_ε is a C^1 function and, for every $x \in \mathbb{R}^n$ and $\rho < \frac{1}{n}$,*

$$n^2\rho^2 < \varphi_\varepsilon(x) < 1 - n^2\rho^2$$

implies

$$\text{dist}(x, \partial E) \leq (1 - \rho)\varepsilon.$$

Proof: Since η_ε is Lipschitz continuous and spt $\eta_\varepsilon \subseteq \bar{B}_\varepsilon$, we obviously have that φ_ε is Lipschitz continuous and hence differentiable almost everywhere. Furthermore

$$D\varphi_\varepsilon(x) = \int D\eta_\varepsilon(x)\varphi_E(x - z)dz$$

and hence

$$|D\varphi_\varepsilon(x_1) - D\varphi_\varepsilon(x_2)| \le C_\varepsilon \int_{|z| < \varepsilon} |\varphi_E(x_1 - z) - \varphi_E(x_2 - z)| dz.$$

When $x_1 \to x_2$ the integral on the right hand side tends to zero, thus proving φ_ε is C^1.

Now suppose $x \in \mathbb{R}^n$ and let $d = \text{dist}(x, E)$. We have

$$\varphi_\varepsilon(x) = \int_E \eta_\varepsilon(x - y) dy =$$

$$= \int_{E - B_d} \eta_\varepsilon(x - y) dy \le n(n + 1)\varepsilon^{-n} \int_d^\varepsilon \left(1 - \frac{t}{\varepsilon}\right) t^{n-1} dt =$$

$$= 1 - n(n + 1)\left(\frac{d}{\varepsilon}\right)^n \left[\frac{1}{n} - \frac{d}{\varepsilon(n + 1)}\right].$$

If $d/\varepsilon \ge 1 - \rho$, then

$$\varphi_\varepsilon(x) \le 1 - (1 - \rho)^n(1 + n\rho) \le 1 - (1 - n\rho)(1 + n\rho) = n\rho^2.$$

Thus, if $\varphi_\varepsilon(x) > n^2\rho^2$, we must have

$$\text{dist}(x, E) < \varepsilon(1 - \rho).$$

In a similar way, if $\varphi_\varepsilon(x) < 1 - n^2\rho^2$, we have

$$\text{dist}(x, \mathbb{R}^n - E) < \varepsilon(1 - \rho)$$

and the conclusion follows immediately.   □

7.2 Lemma: *Suppose $f \in BV(B_1)$, $\tau < 1$, $\varepsilon > 0$ and $\tau + \varepsilon \le 1$. Then*

$$(7.1) \quad \int_{B_\tau} |f_\varepsilon - f| dx \le \varepsilon \int_{B_{\tau + \varepsilon}} |Df|,$$

$$(7.2) \quad \int_{B_\tau} |Df_\varepsilon| - \int_{B_\tau} |Df| \le \int_{B_{\tau + \varepsilon} - B_\tau} |Df|.$$

Proof: As in the proof of Proposition 1.15, we may show

$$\int\limits_{B_\tau} |Df_\varepsilon| \leqq \int\limits_{B_{\tau+\varepsilon}} |Df|$$

and so (7.2) follows at once.

To prove (7.1) we note that if $f \in C^1(B_1)$ then

$$|f_\varepsilon(x) - f(x)| \leqq \int\limits_{B_\varepsilon} \eta_\varepsilon(z) |f(x-z) - f(x)| dz \leqq \int\limits_{B_\varepsilon} |z| \eta_\varepsilon(z) dz \int\limits_0^1 |Df(x-tz)| dt$$

and hence

$$\int\limits_{B_\tau} |f_\varepsilon - f| dx \leqq \int\limits_{B_\tau} \{\varepsilon \int\limits_{B_\varepsilon} \eta_\varepsilon(z) dt \int\limits_0^1 |Df(x-tz)| dt\} dx \leqslant$$

$$\leqq \varepsilon \int\limits_{B_\varepsilon} \eta_\varepsilon(z) dz \int\limits_0^1 dt \int\limits_{B_{\tau+\varepsilon}} |Df(y)| dy = \varepsilon \int\limits_{B_{\tau+\varepsilon}} |Df| dx.$$

(7.1) follows now for any $f \in BV(B_1)$ by approximation. □

The next theorem is important because it shows that, if we have a minimal set whose characteristic function has distributional derivatives lying approximately in one direction, in the sense that

$$\int\limits_B (|D\varphi_E| - D_n\varphi_E) \leqq \gamma,$$

then the mollified functions of the previous two lemmas have derivatives which also lie in approximately the same direction.

7.3 Theorem: *Suppose E is a Caccioppoli set satisfying*

(7.3) $\psi(E, 1) = 0,$

(7.4) $\int\limits_{B_1} (|D\varphi_E| - D_n\varphi_E) \leqq \gamma.$

Then for each positive integer p there exists a constant $\lambda = \lambda(\gamma)$ (depending on γ and p), converging to 0 as γ converges to 0, such that, if $\varepsilon = \gamma^p$ and

$$\varphi(x) = \int \eta_\varepsilon(x-y)\varphi_E(y) dy = (\varphi_E)_\varepsilon,$$

then

$$(7.5) \quad \inf\left\{\frac{D_n\varphi(x)}{|D\varphi(x)|}; \; |x| < 1 - 2\gamma^{\frac{1}{2(n-1)}} \text{ and } n^2\gamma^2 < \varphi(x) < 1 - n^2\gamma^2\right\} > 1 - \lambda(\gamma).$$

Proof: Let $\sigma = \gamma^{\frac{1}{2(n-1)}}$ and suppose $|x| < 1 - 2\sigma$. Since the set over which we take the infimum is obviously empty if $n^2\gamma^2 > \frac{1}{2}$, we may assume $\gamma < 1$ and hence $\varepsilon = \gamma^p < \sigma$. Now we have

$$D_n\varphi(x) = \int \eta_\varepsilon(x - y)D_n\varphi_E(y)$$

$$|D\varphi(x)| \leqq \int \eta_\varepsilon(x - y)|D\varphi_E(y)|.$$

We estimate $|D\varphi(x)| - D_n\varphi(x)$ in terms of $|D\varphi(x)|$ to obtain $\lambda(\gamma)$, and to do this we estimate

$$\int_{B(x,\varepsilon)} \eta_\varepsilon(x - y)(|D\varphi_E(y)| - D_n\varphi_E(y))$$

in terms of

$$\int_{B(x,\varepsilon)} \eta_\varepsilon(x - y)|D\varphi_E(y)|.$$

For simplicity we shall denote $B(x, r)$ by B_r unless we wish to note the dependence on x. The proof is mainly a technical one making use of estimates from the previous two lemmas and also from Chapter 5.

Let $\delta \in (0, \frac{1}{2})$ be a constant which will be chosen later. We estimate $\int \eta_\varepsilon(x - y)(|D\varphi_E(y)| - D_n\varphi_E(y))$ in two parts; firstly in the ball $B_{\varepsilon(1 - 2\delta)}$ and then later in the remainder of B_ε.

By a slight variation of Lemma 2.2 (noting that in this case ρ remains constant), we can prove that there exists a finite number of points $\xi_1, \ldots, \xi_N$ in $\partial^*E \cap B_{\varepsilon(1 - 2\delta)}$ such that

$$B(\xi_i, \delta\varepsilon) \cap B(\xi_j, \delta\varepsilon) = \varnothing \qquad i \neq j$$

$$\partial E \cap B_{\varepsilon(1 - 2\delta)} \subseteq \bigcup_{i=1}^{N} B(\xi_i, 2\delta\varepsilon).$$

Then we have

$$\int_{B_\varepsilon} \eta_\varepsilon(x - y)|D\varphi_E(y)| \geq \sum_{i=1}^{N} \int_{B(\xi_i, \delta\varepsilon)} \eta_\varepsilon(x - y)|D\varphi_E(y)|$$

and

$$\int_{B_{\varepsilon(1-2\delta)}} \eta_\varepsilon(x - y)(|D\varphi_E(y)| - D_n\varphi_E(y)) \leq$$

$$\leq \sum_{i=1}^{N} \int_{B(\xi_i, 2\delta\varepsilon)} \eta_\varepsilon(x - y)(|D\varphi_E(y)| - D_n\varphi_E(y)).$$

So for each i we estimate

$$\int_{B(\xi_i, 2\delta\varepsilon)} \eta_\varepsilon(x - y)(|D\varphi_E(y)| - D_n\varphi_E(y))$$

in terms of

$$\int_{B(\xi_i, \delta\varepsilon)} \eta_\varepsilon(x - y)|D\varphi_E(y)|.$$

Now in the ball $B(\xi_i, \delta\varepsilon)$ the function $\eta_\varepsilon(x - y)$ is bounded below by

$\dfrac{n + 1}{\omega_n} \varepsilon^{-n}(1 - \delta - \dfrac{|x - \xi_i|}{\varepsilon}) > 0$ and so on

$$\int_{B(\xi_i, \delta\varepsilon)} \eta_\varepsilon(x - y)|D\varphi_E(y)| \geq \frac{n + 1}{\omega_n} \varepsilon^{-n}\left(1 - \delta - \frac{|x - \xi_i|}{\varepsilon}\right) \int_{B(\xi_i, \delta\varepsilon)} |D\varphi_E(y)|.$$

Furthermore, we have that, by (5.16).

$$\int_{B(\xi_i, \delta\varepsilon)} |D\varphi_E(y)| \geq \omega_{n-1}(\delta\varepsilon)^{n-1}.$$

Hence

$$(7.6) \quad \int_{B(\xi_i, \delta\varepsilon)} \eta_\varepsilon(x - y)|D\varphi_E(y)| \geq \omega_{n-1}\frac{n + 1}{\omega_n}\delta^{n-1}\varepsilon^{-1}\left(1 - \delta - \frac{|x - \xi_i|}{\varepsilon}\right).$$

On the other hand, by estimating $\eta_\varepsilon(x - y)$ above, it follows that

$$\int_{B(\xi_i, 2\delta\varepsilon)} \eta_\varepsilon(x - y)(|D\varphi_E(y)| - D_n\varphi_E(y)) \leqslant$$

$$\leqq \frac{n + 1}{\omega_n} \varepsilon^{-n}\left(1 + 2\delta - \frac{|x - \xi_i|}{\varepsilon}\right) \int_{B(\xi_i, 2\delta\varepsilon)} (|D\varphi_E(y)| - D_n\varphi_E(y)).$$

Now using (5.8) and the fact that $\sigma > 2\delta\varepsilon$, we obtain

$$\int_{B(\xi_i, 2\delta\varepsilon)} (|D\varphi_E(y)| - D_n\varphi_E(y)) \leqq (2\delta\varepsilon)^{n-1}\{\sigma^{1-n} \int_{B(\xi_i, \sigma)} (|D\varphi_E(y)| - D_n\varphi_E(y)) +$$

$$+ \sigma^{1-n}\int_{B(\xi_i, \sigma)} D_n\varphi_E(y) - (2\delta\varepsilon)^{1-n}\int_{B(\xi_i, 2\delta\varepsilon)} D_n\varphi_E(y)\}$$

and from (5.15), (7.4) and the assumption $|x| < 1 - 2\sigma$, we have

$$\int_{B(\xi_i, 2\delta\varepsilon)} (|D\varphi_E(y)| - D_n\varphi_E(y)) \leqq$$

$$\leqq (2\delta\varepsilon)^{n-1}\left\{\gamma^{1/2} + \sqrt{n\omega_n}\sqrt{1 + (n-1)\log\frac{\sigma}{2\delta\varepsilon}}\left(\sigma^{1-n}\int_{B(\xi_i, \sigma)} |D\varphi_E(y)| - \right.\right.$$

$$\left.\left. - (2\delta\varepsilon)^{1-n}\int_{B(\xi_i, 2\delta\varepsilon)} |D\varphi_E(y)|\right)^{1/2}\right\}.$$

Finally we note that

$$\sigma^{1-n}\int_{B(\xi_i, \sigma)} |D\varphi_E(y)| - (2\delta\varepsilon)^{1-n}\int_{B(\xi_i, 2\delta\varepsilon)} |D\varphi_E(y)| =$$

$$= \sigma^{1-n}\int_{B(\xi_i, \sigma)} (|D\varphi_E(y)| - D_n\varphi_E) +$$

$$+ \sigma^{1-n}\int_{B(\xi_i, \sigma)} D_n\varphi_E(y) - (2\delta\varepsilon)^{1-n}\int_{B(\xi_i, 2\delta\varepsilon)} |D\varphi_E(y)|$$

The Gauss-Green formula gives

$$\sigma^{1-n}\int_{B(\xi_i, \sigma)} D_n\varphi_E(y) \leqq \omega_{n-1},$$

and $\xi_i \in \partial^* E$ implies

$$(2\delta\varepsilon)^{1-n}\int_{B(\xi_i,2\delta\varepsilon)}|D\varphi_E(y)|\geqq \omega_{n-1},$$

whence applying inequality (7.4) we obtain

$$\int_{B(\xi_i,2\delta\varepsilon)}\eta_\varepsilon(x-y)(|D\varphi_E(y)|-D_n\varphi_E(y))\leqslant$$

$$\leqq \frac{n+1}{\omega_n}2^{n-1}\delta^{n-1}\varepsilon^{-1}\left(1+2\delta-\frac{|x-\xi_i|}{\varepsilon}\right)$$

$$\left\{\gamma^{1/2}+\sqrt{n\omega_n\left[1+(n-1)\log\frac{\sigma}{2\delta\varepsilon}\right]}\gamma^{1/4}\right\}.$$

Comparing this with (7.6) and observing that

$$\frac{1+2\delta-\dfrac{|x-\xi_i|}{\varepsilon}}{1-\delta-\dfrac{|x-\xi_i|}{\varepsilon}}=1+\frac{3\delta}{1-\dfrac{|x-\xi_i|}{\varepsilon}-\delta}\leqq 1+\frac{3\delta}{\delta}=4,$$

we have

$$\int_{B(\xi_i,2\delta\varepsilon)}\eta_\varepsilon(x-y)(|D\varphi_E(y)|-D_n\varphi_E(y))\leqslant$$

$$\leqq \frac{2^{n+1}}{\omega_{n-1}}\gamma^{1/4}\left\{\gamma^{1/4}+\sqrt{n\omega_n\left[1+(n-1)\log\frac{\sigma}{2\delta\varepsilon}\right]}\right\}$$

$$\int_{B(\xi_i,\delta\varepsilon)}\eta_\varepsilon(x-y)|D\varphi_E(y)|.$$

Hence summing from $i=1$ to N, we obtain

$$(7.7)\quad \int_{B_{\varepsilon(1-2\delta)}}\eta_\varepsilon(x-y)(|D\varphi_E(y)|-D_n\varphi_E(y))\leqslant$$

$$\leqq \frac{2^{n+1}}{\omega_{n-1}}\gamma^{1/4}\left\{\gamma^{1/4}+\sqrt{n\omega_n\left[1+(n-1)\log\frac{\sigma}{2\delta\varepsilon}\right]}\right\}$$

$$\int_{B_\varepsilon} \eta_\varepsilon(x - y)|D\varphi_E(y)|.$$

We now perform a similar estimate for the integral over the set $C = B_\varepsilon - B_{\varepsilon(1-2\delta)}$. In the set C, $\eta_\varepsilon(x - y)$ is bounded above by $\frac{n+1}{\omega_n} \varepsilon^{-n} 2\delta$ and hence

$$(7.8) \quad \int_C \eta_\varepsilon(x - y)(|D\varphi_E(y)| - D_n\varphi_E(y)) \leq 2 \int_C \eta_\varepsilon(x - y)|D\varphi_E(y)| \leq$$

$$\leq 4 \frac{(n+1)}{\omega_n} \varepsilon^{-n}\delta \int_{B_\varepsilon} |D\varphi_E(y)| \leq$$

$$\leq 2n(n+1)\varepsilon^{-1}\delta, \text{ by (5.14)}.$$

So far we have used only $|x| < 1 - 2\sigma$, but now we introduce the hypothesis

$$n^2\gamma^2 < \varphi(x) < 1 - n^2\gamma^2.$$

From Lemma 7.1 we conclude that there exists a $\xi \in \partial^* E$ with $|\xi - x| \leq (1 - \gamma)\varepsilon$. Then

$$B\left(\xi, \frac{\gamma\varepsilon}{2}\right) \subseteq B\left(x, \left(1 - \frac{\gamma}{2}\right)\varepsilon\right)$$

and hence $\eta_\varepsilon(x - y)$ is bounded below by $\dfrac{n+1}{\omega_n} \varepsilon^{-n}\dfrac{\gamma}{2}$ in the set $B(\xi, \frac{\gamma\varepsilon}{2})$. It now follows that

$$\int_{B_\varepsilon} \eta_\varepsilon(x - y)|D\varphi_E(y)| \geq \int_{B\left(\xi, \frac{\gamma\varepsilon}{2}\right)} \eta_\varepsilon(x - y)|D\varphi_E(y)| \geq$$

$$\geq \frac{n+1}{\omega_n} \frac{\omega_{n-1}}{2^n} \gamma^{n-1} \varepsilon^{-1}$$

and, comparing this with (7.8),

$$\int_C \eta_\varepsilon(x - y)(|D\varphi_E(y)| - D_n\varphi_E(y)) \leq$$

$$\leq 2^{n+1} \frac{n\omega_n}{\omega_{n-1}} \delta\gamma^{1-n} \int_{B_\varepsilon} \eta_\varepsilon(x - y)|D\varphi_E(y)|.$$

Adding this to (7.7) we obtain:

$$(7.9) \quad \int_{B_\varepsilon} \eta_\varepsilon(x - y)(|D\varphi_E(y)| - D_n\varphi_E(y)) \leqq \lambda(\gamma) \int_{B_\varepsilon} \eta_\varepsilon(x - y)|D\varphi_E(y)|$$

with

$$\lambda(\gamma) = 2^{n+1} \frac{n\omega_2}{\omega_{n-1}} \delta\gamma^{1-n} + \frac{2^{n+1}}{\omega_{n-1}} \gamma^{1/4} \left\{ \gamma^{1/4} + \sqrt{n\omega_n \left(1 + (n-1)\log \frac{\sigma}{2\delta\varepsilon} \right)} \right\}$$

The choice $\delta = \gamma^n$ gives the required properties for $\lambda(\gamma)$. $\qquad\qquad\square$

7.4 Remark: The last integral in (7.9) is obviously positive for x satisfying $|x| < 1 - 2\sigma$, $n^2\gamma^2 \leqq \varphi(x) \leqq 1 - n^2\gamma^2$, and therefore for such x we have:

$$D_n\varphi(x) = \int_{B_\varepsilon} \eta_\varepsilon(x - y)D_n\varphi_E(y) \geqq [1 - \lambda(\gamma)] \int_{B_\varepsilon} \eta_\varepsilon(x - y)|D\varphi_E(y)| > 0.$$

In particular, the level sets of φ:

$$S(\vartheta) = \{x \in B : |x| < 1 - 2\sigma, \ \varphi(x) > \vartheta\}$$

have C^1-boundaries for every ϑ, $n^2\gamma^2 < \vartheta < 1 - n^2\gamma^2$.

The aim of this chapter is to prove a De Giorgi type lemma for sequences of Caccioppoli sets. To do this we approximate by C^1 sets and then use the results of Chapter 6. Hence it is necessary to show that our approximating sequence satisfies the conditions required by Chapter 6. We are now in a position to do this using the previous lemmas and it is then only a short step to prove the required result.

7.5 Lemma: *Suppose $\{E_j\}$ is a sequence of Caccioppoli sets such that*

$$(7.10) \quad \psi(E_j, 1) = 0,$$

$$(7.11) \quad \int_{B_1} |D\varphi_{E_j}| - \int_{B_1} D_n\varphi_{E_j} \leqq \gamma_j,$$

$$(7.12) \quad \sum_{j=1}^{\infty} \gamma_j < \infty.$$

Then for almost every $t \in (0, 1)$ *there exists a sequence of sets* $\{L_j\}$ *such that*

(7.13) $\lim\limits_{j \to \infty} \gamma_j^{-1} \psi(L_j, t) = 0,$

(7.14) $\lim\limits_{j \to \infty} \gamma_j^{-1} \left\{ \int_{B_t} |D\varphi_{L_j}| - \int_{B_t} |D\varphi_{E_j}| \right\} = 0,$

(7.15) $\lim\limits_{j \to \infty} \gamma_j^{-1} \left| \int_{B_t} D\varphi_{L_j} - \int_{B_t} D\varphi_{E_j} \right| = 0,$

(7.16) $\partial L_j \cap B_t$ *is a* C^1 *hypersurface,*

and for every $s < t$

(7.17) $\lim\limits_{j \to \infty} \inf \{v_n^{(j)}(x) : |x| < s, \ x \in \partial L_j\} = 1,$

where $v^{(j)}(x)$ *is the normal to* ∂L_j *at* x.

Proof: Let $\varepsilon_j = \gamma_j^4$ and $f_j(x) = \int \eta_{\varepsilon_j}(x - y)\varphi_{E_j}(y)dy$. Then from (7.1) and (5.14), if $\tau < 1$,

(7.18) $\lim\limits_{j \to \infty} \sup \gamma_j^{-4} \int_{B_\tau} |f_j - \varphi_{E_j}| dx \leq \frac{1}{2} n\omega_n$

and hence as $\gamma_j \to 0$ there exists a null set $N_1 \subseteq (0, 1)$ such that for $t \in (0, 1) - N_1$

(7.19) $\lim\limits_{j \to \infty} \gamma_j^{-3} \int_{\partial B_t} |f_j - \varphi_{E_j}| dH_{n-1} = 0.$

Now let μ denote the measure $\sum\limits_{j=1}^{\infty} \gamma_j |D\varphi_{E_j}|$. From (7.12) and (5.14),

$$\int_{B_1} d\mu < \infty$$

and so, as the function $\int_{B_t} d\mu$ is monotone increasing in t and hence differentiable almost everywhere, we may conclude that there exists a null set N_2 such that for every $t \in (0, 1) - N_2$

$$\lim_{j \to \infty} \gamma_j^{-4} \int_{B_{t + \gamma_j^4} - B_t} d\mu < \infty.$$

Thus, by definition of μ,

$$\lim_{j \to \infty} \sup \gamma_j^{-3} \int_{B_{t - \gamma_j^4} - B_t} |D\varphi_{E_j}| < \infty \qquad \text{for } t \in (0, 1) - N_2$$

and so, from (7.2) and the fact that $\gamma_j \to 0$,

$$(7.20) \quad \lim_{j \to \infty} \sup \gamma_j^{-2} \left\{ \int_{B_t} |Df_j| - \int_{B_t} |D\varphi_{E_j}| \right\} \leq 0$$

for $t \in (0, 1) - N_2$.

If we let $S_j(\vartheta) = \{ x \in B_1 : f_j(x) > \vartheta \}$ then from Theorem 1.23 we have

$$\int_{B_t} |Df_j| = \int_0^1 d\vartheta \int_{B_t} |D\varphi_{S_j(\vartheta)}| \geq \int_{n^2 \gamma_j^2}^{1 - n^2 \gamma_j^2} d\vartheta \int_{B_t} |D\varphi_{S_j(\vartheta)}|.$$

Hence it follows that there exists a number $\vartheta_j \in (n^2 \gamma_j^2, 1 - n^2 \gamma_j^2)$ such that

$$(7.21) \quad (1 - 2n^2 \gamma_j^2) \int_{B_t} |D\varphi_{S_j(\vartheta_j)}| \leq \int_{B_t} |Df_j|$$

Moreover, by Remark 7.4, the boundary of $L_j = S_j(\vartheta_j)$ is regular. From (7.20) and (7.21)

$$(7.22) \quad \lim_{j \to \infty} \sup \gamma_j^{-1} \left\{ \int_{B_t} |D\varphi_{L_j}| - \int_{B_t} |D\varphi_{E_j}| \right\} \leq 0 \qquad \text{for } t \in (0, 1) - N_2.$$

On the other hand from Lemma 1.25

$$\int_{\partial B_t} |\varphi_{L_j} - \varphi_{E_j}| dH_{n-1} \leq n^{-2} \gamma_j^{-2} \int_{\partial B_t} |f_j - \varphi_{E_j}| \qquad \text{for } t \in (0, 1)$$

and hence from (7.19)

$$(7.23) \quad \lim_{j \to \infty} \gamma_j^{-1} \int_{\partial B_t} |\varphi_{L_j} - \varphi_{E_j}| dH_{n-1} = 0 \qquad \text{for } t \in (0, 1) - N_1.$$

From (7.23) and inequality (5.5) of Remark 5.7 we conclude that

$$(7.24) \quad \limsup_{j \to \infty} \gamma_j^{-1} \left\{ \int_{B_t} |D\varphi_{E_j}| - \int_{B_t} |D\varphi_{L_j}| \right\} \leqq 0 \qquad \text{for } t \in (0, 1) - N_1$$

which together with (7.22) gives (7.14).

On the other hand, if $f \in BV(B_1)$, then by (2.14)

$$\int_{B_t} Df = \int_{\partial B_t} f \frac{x}{|x|} dH_{n-1}$$

and so (7.15) follows from (7.23).

Finally, (7.13) is a consequence of (7.23), (7.14) and Lemma 5.6 and (7.16), (7.17) follow from Theorem 7.3.   □

7.6   Theorem: *Suppose $\{E_j\}$ is a sequence of Caccioppoli sets such that*

$$\psi(E_j, 1) = 0,$$

$$\int_{B_1} |D\varphi_{E_j}| - \int_{B_1} D_n \varphi_{E_j} \leqq \gamma_j,$$

$$\sum_{j=1}^{\infty} \gamma_j < \infty.$$

Then for every α, $0 < \alpha < 1$, we have

$$(7.25) \quad \limsup_{j \to \infty} \gamma_j^{-1} \left\{ \int_{B_\alpha} |D\varphi_{E_j}| - | \int_{B_\alpha} D\varphi_{E_j} | \right\} \leqq \alpha^{n+1}.$$

Proof: Let $\{L_j\}$ be the sequence of sets determined by Lemma 7.5 and suppose that the inequalities of Lemma 7.5 hold at $t < 1$. Let s be a positive number smaller than t. Then the sequence $\{L_j\}$ satisfies the hypotheses of Lemma 6.4 with

$\rho = s$ and sequence β_j such that $\lim_{j \to \infty} \beta_j / \gamma_j = 1$. From Lemma 6.4, (7.14) and (7.17) we have immediately

$$\limsup_{j \to \infty} \gamma_j^{-1} \left\{ \int_{B_\alpha} |D\varphi_{E_j}| - | \int_{B_\alpha} D\varphi_{E_j} | \right\} \leqq \left(\frac{\alpha}{s} \right)^{n+1}.$$

Now choosing s and t close to 1, (7.25) follows.   □

8. Regularity of Minimal Surfaces

In this chapter we can finally prove that partial regularity of minimal surfaces; namely we show that the reduced boundary $\partial^* E$ is analytic and the only possible singularities must occur in $\partial E - \partial^* E$. Our major task in the following chapters will be to obtain an estimate for the size of $\partial E - \partial^* E$. As mentioned before, the main step in the regularity theory is the De Giorgi lemma. We show that a minimal surface is regular at points which satisfy the hypotheses of the lemma. Obviously by the definition of reduced boundary this must include all the points in $\partial^* E$ and we show that the converse also holds and further that $\partial^* E$ is relatively open in ∂E.

8.1 Theorem: *For every $n \geq 2$ and every α, $0 < \alpha < 1$, there exists a constant $\sigma(n, \alpha)$ such that: if E is a Caccioppoli set in $\mathbb{R}^n$, $x \in \mathbb{R}^n$, $\rho > 0$ and*

(8.1) $\psi(E, B(x, \rho)) = 0,$

(8.2) $\left\{ \int\limits_{B(x,\rho)} |D\varphi_E| - \left| \int\limits_{B(x,\rho)} D\varphi_E \right| \right\} < \sigma(n, \alpha)\rho^{n-1},$

then

(8.3) $\left\{ \int\limits_{B(x,\alpha\rho)} |D\varphi_E| - \left| \int\limits_{B(x,\alpha\rho)} D\varphi_E \right| \right\} \leq \alpha^n \left\{ \int\limits_{B(x,\rho)} |D\varphi_E| - \left| \int\limits_{B(x,\rho)} D\varphi_E \right| \right\}.$

Proof: Suppose the theorem is not true. Then there exist $n \geq 2$, $0 < \alpha < 1$, a sequence $\{F_j\}$ of Caccioppoli sets in $\mathbb{R}^n$, a sequence $\{x_j\}$ in $\mathbb{R}^n$, a sequence $\{\rho_j\}$ in $\mathbb{R}$ and a sequence $\{\gamma_j\}$ in $\mathbb{R}$ such that

$\psi(F_j, B(x_j, \rho_j)) = 0,$

$\int\limits_{B(x_j,\rho_j)} |D\varphi_{F_j}| - \left| \int\limits_{B(x_j,\rho_j)} D\varphi_{F_j} \right| = \gamma_j \rho_j^{n-1},$

$\sum\limits_{j=1}^{\infty} \gamma_j < \infty$

and

$$\int_{B(x_j,\alpha\rho_j)} |D\varphi_{F_j}| - \Big|\int_{B(x_j,\alpha\rho_j)} D\varphi_{F_j}\Big| > \alpha^n \gamma_j \rho_j^{n-1}.$$

For each j, we apply a translation to F_j, taking x_j to the origin, then a rotation taking the vector $\int_{B(0,\rho_j)} D\varphi_{F_j}$ onto the x_n-axis and finally a dilation of ratio ρ_j. Call the resulting set E_j. Then

$$\psi(E_j, 1) = 0,$$

$$\int_{B_1} |D\varphi_{E_j}| - \int_{B_1} D_n\varphi_{E_j} = \gamma_j,$$

$$\int_{B_\alpha} |D\varphi_{E_j}| - \Big|\int_{B_\alpha} D\varphi_{E_j}\Big| > \alpha^n \gamma_j.$$

The sequence $\{E_j\}$ must then contradict Theorem 7.6 and the theorem is proved. $\square$

The fact that points x, satisfying the hypotheses of Theorem 8.1, are in $\partial^* E$ will be an easy consequence of the following.

8.2 Theorem: *Suppose the hypotheses of Theorem 8.1 hold and $x \in \partial E$. Then for every s, t such that $0 < s < t < \rho$*

$$(8.4) \quad |v_s(x) - v_t(x)| \leq \eta(\alpha, n)\sqrt{\frac{t}{\rho}},$$

where

$$\eta(\alpha, n) = \frac{(2 - \sqrt{\alpha})}{(1 - \sqrt{\alpha})}\sqrt{\frac{\sigma(n, \alpha)}{\omega_{n-1}\alpha^n}}$$

Proof: By appropriate transformations we may suppose $x = 0$ and $\rho = 1$. We begin with the special case

$$t = \alpha^j, \ s = \beta\alpha^j \text{ for some } j \text{ and } \alpha \leq \beta < 1.$$

Let

$$u_j = v_{\alpha^j} = \frac{\int\limits_{B_{\alpha^j}} D\varphi_E}{\int\limits_{B_{\alpha^j}} |D\varphi_E|}$$

$$v_j = v_{\beta_{\alpha^j}}, \; m_j = \int\limits_{B_{\alpha^j}} |D\varphi_E| \text{ and } \mu_j = \int\limits_{B_{\beta\alpha^j}} |D\varphi_E|.$$

Since $|u_j| \leq 1$ and $|v_j| \leq 1$ we have

$$|u_j - v_j| \leq \sqrt{2}\sqrt{1 - (u_j, v_j)}.$$

Now

$$(1 - (u_j, v_j))\mu_j = \int\limits_{B_{\beta\alpha^j}} (|D\varphi_E| - \langle u_j, D\varphi_E \rangle) \leq$$

$$\leq \int\limits_{B_{\beta\alpha^j}} (|D\varphi_E| - \langle u_j, D\varphi_E \rangle) =$$

$$= m_j(1 - |u_j|^2) \leq 2m_j(1 - |u_j|).$$

On the other hand from (8.3) and (8.2), repeating j times,

$$m_j(1 - |u_j|) \leq \alpha^{nj}\sigma(n, \alpha)$$

and so

$$|u_j - v_j| \leq 2\sqrt{\sigma(n, \alpha)}\left(\frac{\alpha^{nj}}{\mu_j}\right)^{1/2}.$$

However, as we are assuming that $0 \in \partial E$, by (5.16) we have

$$\mu_j \geq \omega_{n-1}(\alpha^j\beta)^{n-1} \geq \omega_{n-1}\alpha^{(j+1)(n-1)}$$

and so

$$|u_j - v_j| \leq 2\sqrt{\frac{\sigma(n, \alpha)}{\omega_{n-1}\alpha^n}}\sqrt{\alpha^{j+1}}.$$

Now consider the case of general s and t with $0 < s < t < 1$. Let j and k be the two integers such that

$$\alpha^{j+1} < t \leq \alpha^j \quad \text{and} \quad \alpha^{j+k} \leq s < \alpha^{j+k-1}.$$

We have

$$|v_t - v_s| \leq |v_t - u_j| + \sum_{i=0}^{k-2} |u_{j+i} - u_{j+i+1}| + |u_{j+k-1} - v_s| \leq$$

$$\leq 2 \sqrt{\frac{\sigma(n, \alpha)}{\omega_{n-1}\alpha^n}} \left\{ \sqrt{\alpha^{j+1}} + \sum_{i=0}^{\infty} \sqrt{\alpha^{j+i+1}} \right\} =$$

$$= \sqrt{\frac{\sigma(n, \alpha)}{\omega_{n-1}\alpha^n}} \sqrt{\alpha^{j+1}} \left\{ 1 + \sum_{i=0}^{\infty} \alpha^{i/2} \right\} =$$

$$= \frac{(2 - \sqrt{\alpha})}{(1 - \sqrt{\alpha})} \sqrt{\frac{\sigma(n, \alpha)}{\omega_{n-1}\alpha^n}} \sqrt{\alpha^{j+1}} \leq$$

$$\leq \frac{(2 - \sqrt{\alpha})}{(1 - \sqrt{\alpha})} \sqrt{\frac{\sigma(n, \alpha)}{\omega_{n-1}\alpha^n}} \sqrt{t}. \qquad \square$$

8.3   Corollary: *Suppose x satisfies the hypotheses of Theorem 8.2 Then $x \in \partial^* E$ and*

$$|v(x) - v_t(x)| \leq \eta(n, \alpha) \sqrt{\frac{t}{\rho}}.$$

Finally we prove the regularity of $\partial^* E$ together with the fact that $\partial^* E$ is open in ∂E. These facts follow easily by some standard results from the theory of elliptic partial differential equations once we can prove that $\partial^* E$ is a C^1-surface. To do this we need only show that $v(x)$ is continuous.

8.4   Theorem: *Suppose E is a Caccioppoli set in $\mathbb{R}^n$, $x \in \partial E$, $\rho > 0$ and $0 < \alpha < 1$ are such that*

$$\psi(E, B(x, \rho)) = 0,$$

$$\int\limits_{B(x,\rho)} |D\varphi_E| - | \int\limits_{B(x,\rho)} D\varphi_E| \leq \sigma(n,\alpha)\rho^{n-1}.$$

Then $\partial E \cap B(x,r)$ is an analytic hypersurface for $r = \rho(\alpha - \alpha^{\frac{n}{n-1}})$.

Proof: Let $z \in \partial E \cap B(x,r)$ and $R = \rho\alpha^{\frac{n}{n-1}}$. As $B(z,R) \subseteq B(x,\alpha\rho)$, we have

$$\int\limits_{B(z,R)} |D\varphi_E| - | \int\limits_{B(z,R)} D\varphi_E| \leq \int\limits_{B(x,\alpha\rho)} |D\varphi_E| - | \int\limits_{B(x,\alpha\rho)} D\varphi_E| \leq \sigma(n,\alpha)\alpha^n\rho^{n-1}$$

and hence

$$\int\limits_{B(z,R)} |D\varphi_E| - | \int\limits_{B(z,R)} D\varphi_E| \leq \sigma(n,\alpha)R^{n-1}.$$

Thus z satisfies the hypotheses of Corollary 8.3 and so $z \in \partial^* E$ and

$$(8.5) \qquad \left| v(z) - v_t(z) \right| = \left| v(z) - \frac{\int\limits_{B(z,t)} D\varphi_E}{\int\limits_{B(z,t)} |D\varphi_E|} \right| \leq \eta(n,\alpha)\sqrt{\frac{t}{R}}.$$

(8.5) must hold for every $z \in B(x,r) \cap \partial E$. From (8.5), by a simple integration we obtain

$$(8.6) \qquad \left| v(z) - \frac{\int\limits_0^t ds \int\limits_{B(z,s)} D\varphi_E}{\int\limits_0^t ds \int\limits_{B(z,s)} |D\varphi_E|} \right| \leq \eta(n,\alpha)\sqrt{\frac{t}{R}}.$$

The function

$$\tilde{v}(t,z) = \frac{\int\limits_0^t ds \int\limits_{B(z,s)} D\varphi_E}{\int\limits_0^t ds \int\limits_{B(z,s)} |D\varphi_E|}$$

is continuous in t and z and since from (8.6)

$$\lim_{t \to 0} \tilde{v}(t,z) = v(z) \text{ uniformly for } z \in B(x,r),$$

we must have that $v(z)$ is continuous on $\partial E \cap B(x, r)$ and hence, by Theorem 4.11, $\partial E \cap B(x, r)$ is a C^1-hypersurface.

The analyticity of $\partial E \cap B(x, r)$ follows from the regularity theory for elliptic partial differential equations (see [MCB1]).   ☐

We have proved ∂E is analytic in a neighbourhood of every point in $\partial^* E$. It is possible that the singular set $\partial E - \partial^* E$ may be nonempty and indeed examples where this is true may be constructed. However we have

8.5 Theorem: *Suppose E is a Caccioppoli set satisfying*

$$\psi(E, 1) = 0.$$

Then

(8.7) $H_{n-1}(\partial E - \partial^* E) = 0.$

Proof: Let K be any compact set contained in $\partial E - \partial^* E$. Then, by Definition 3.3 and the note following it,

$$\int_K |D\varphi_E| = 0.$$

Hence, as $|D\varphi_E|$ is a Radon measure, for every $\varepsilon > 0$ there exists an open set A_ε with $K \subset A_\varepsilon \subset B_1$ such that

$$\int_{A_\varepsilon} |D\varphi_E| < \varepsilon.$$

Let $\eta > 0$. For every $x \in K$ there exists $\rho < \eta$ such that $B(x, 3\rho) \subseteq A_\varepsilon$. Then by Lemma 2.2 we can choose a sequence $\{x_j\}$ such that

$$B(x_j, \rho_j) \cap B(x_i, \rho_i) = \varnothing \qquad \text{if} \quad i \neq j$$

$$\bigcup_{j=1}^{\infty} B(x_j, 3\rho_j) \supseteq K.$$

We now have

$$\sum_{j=1}^{\infty} \int_{B(x_j,\rho_j)} |D\varphi_E| \leq \int_{A_\varepsilon} |D\varphi_E| < \varepsilon.$$

On the other hand by (5.16)

$$\int_{B(x_j,\rho_j)} |D\varphi_E| \geq \omega_{n-1}\rho_j^{n-1}$$

and hence

$$\omega_{n-1} \sum_{j=1}^{\infty} \rho_j^{n-1} \leq \varepsilon.$$

Thus $H_{n-1}(K) \leq 3^{n-1}\varepsilon$ and so as ε is arbitrary $H_{n-1}(K) = 0$ implying (8.7) holds. $\qquad\qquad\square$

9. Minimal Cones

The results of the previous chapter give an upper bound for the dimension of the singularities of minimal surfaces, but leave open the problem of their regularity. As we shall see, the answer to this question depends essentially on the existence or non-existence of minimal cones that are not hyperplanes.

The way to construct these minimal cones is as follows: If E is a minimal set in B_1 and $0 \in \partial E$, then we expand the set E around 0, that is, we consider the sets

$$E_t = \{x \in \mathbb{R}^n : tx \in E\}$$

as $t \to 0$. Each set E_t is minimal and by selecting a subsequence $\{t_n\}$, converging to zero, we can show that E_{t_n} converges to a set C. This set C must be minimal in $\mathbb{R}^n$ and it must be a cone, that is, we may write

$$C = \{tx : t \geq 0 \quad \text{and} \quad x \in A\}$$

for some set A. C is, in some sense, a tangent cone to ∂E at 0. Furthermore we show that ∂E is regular at 0 if and only if C is a half space. Thus the set E can only have singularities if there exist minimal cones in $\mathbb{R}^n$ which have singularities.

We shall be looking at sequences of minimal sets and so many of the results from previous chapters will be useful. In addition we need the following

9.1 Lemma: *Let Ω be an open set in $\mathbb{R}^n$, and let $\{E_j\}$ be a sequence of Caccioppoli sets of least area in Ω, i.e. such that*

$$\psi(E_j, A) = 0 \qquad \forall A \subset\subset \Omega$$

(see Definition 5.1). Suppose that there exists a set E such that

$$\varphi_{E_j} \to \varphi_E \qquad \text{in} \quad L^1_{\text{loc}}(\Omega).$$

Then E has least perimeter in Ω, namely:

(9.1) $\psi(E, A) = 0 \qquad \forall A \subset\subset \Omega.$

Moreover, if $L \subset\subset \Omega$ is any open set such that

$$\int\limits_{\partial L} |D\varphi_E| = 0$$

we have

(9.2) $$\lim_{j \to \infty} \int_L |D\varphi_{E_j}| = \int_L |D\varphi_E|.$$

Proof: Let $A \subset\subset \Omega$. We may suppose that ∂A is smooth, so that for every j:

(9.3) $$\int\limits_A |D\varphi_{E_j}| \leqq H_{n-1}(\partial A),$$

the same inequality holding for E by Theorem 1.9.

For $t > 0$, let

$$A_t = \{x \in \Omega: \text{dist}(x, A) < t\}.$$

We have

$$\lim_{j \to \infty} \int_{A_t} |\varphi_{E_j} - \varphi_E| dx = 0$$

and therefore there exists a subsequence $\{E_{k_j}\}$ such that for almost every t close to 0:

$$\lim_{j \to \infty} \int_{\partial A_t} |\varphi_{E_{k_j}} - \varphi_E| dH_{n-1} = 0.$$

From Lemma 5.6 we have for these t:

$$\lim_{j \to \infty} v(E_{k_j}, A_t) = v(E, A_t)$$

and therefore from Theorem 1.9:

$$\psi(E, A_t) = 0.$$

This proves (9.1). Now let $L \subset\subset \Omega$ be such that $\int\limits_{\partial L} |D\varphi_E| = 0$, and let A be a smooth open set such that $L \subset\subset A \subset\subset \Omega$. Let $\{F_j\}$ be any subsequence of $\{E_j\}$. Reasoning as above we can find a set A_t and a subsequence $\{F_{kj}\}$ such that

$$\lim_{j\to\infty} v(F_{k_j}, A_t) = v(E, A_t).$$

Since $\psi(F_{k_j}, A_t) = \psi(E, A_t) = 0$ we have

$$\lim_{j\to\infty} \int_{A_t} |D\varphi_{F_{k_j}}| = \int_{A_t} |D\varphi_E|,$$

and hence from Proposition 1.13 (with A_t playing the role of Ω):

$$\int_L |D\varphi_E| = \lim_{j\to\infty} \int_L |D\varphi_{F_{k_j}}|$$

proving (9.2).   □

9.2   Remark: Given a sequence of Caccioppoli sets with least perimeter in Ω, it is always possible to select a subsequence, converging in $L^1_{loc}(\Omega)$. Actually, for each open set $A \subset\subset \Omega$ such a subsequence exists by (9.3) and Rellich's Theorem 1.19; a diagonal process gives the required sequence converging in $L^1_{loc}(\Omega)$.

We now consider the case where E is a minimal set and $0 \in \partial E$. We "blow up" E about 0 and show that the expanded sets approach a limit C and that the set C is a cone. Furthermore C itself is a minimal set.

9.3   Theorem: *Suppose E is a minimal set in B_1 (that is $\psi(E, 1) = 0$) such that $0 \in \partial E$. For $t > 0$, let*

$$E_t = \{x \in \mathbb{R}^n : tx \in E\}.$$

Then for every sequence $\{t_j\}$ tending to zero there exists a subsequence $\{s_j\}$ such that E_{s_j} converges locally in $\mathbb{R}^n$ to a set C. Moreover C is a minimal cone.

Proof: Let $t_j \to 0$. We show first that for every $R > 0$ there exists a subsequence $\{\sigma_j\}$ such that E_{σ_j} converges in B_R. We have

$$\int_{B_R} |D\varphi_{E_t}| = t^{1-n} \int_{B_{Rt}} |D\varphi_E|$$

and so choosing t sufficiently small (so that $Rt < 1$) we have that E_t is minimal in B_R and

(9.4) $\quad \int_{B_R} |D\varphi_{E_t}| = t^{1-n} \int_{B_{Rt}} |D\varphi_E| \leqq \dfrac{1}{2} n \omega_n R^{n-1}.$

Hence, by the compactness Theorem 1.19, a subsequence E_{σ_j} will converge to a set C_R in B_R. Taking a sequence $R_i \to \infty$ we obtain, by means of a diagonal process, a set $C \subseteq \mathbb{R}^n$ and a sequence $\{s_j\}$ such that $E_{s_j} \to C$ locally. Now, applying Lemma 9.1, we see that C is minimal and it remains only to show that C is a cone.

Also by Lemma 9.1 we have that, for almost all $R > 0$,

$$\int_{B_R} |D\varphi_{E_{s_j}}| \to \int_{B_R} |D\varphi_C|.$$

Hence if we define

$$p(t) = t^{1-n} \int_{B_t} |D\varphi_E| = \int_{B_1} |D\varphi_{E_t}|,$$

we have, for almost all $R > 0$,

(9.5) $\quad \lim_{j \to \infty} p(s_j R) = R^{1-n} \int_{B_R} |D\varphi_C|.$

Note as well that, by (5.12), $p(t)$ is increasing in t.

If $\rho < R$, then for every j there exists an $m_j > 0$ such that

$$s_j \rho > s_{j+m_j} R.$$

Then

$$p(s_{j+m_j} R) \leqq p(s_j \rho) \leqq p(s_j R)$$

so that

$$\lim_{j \to \infty} p(s_j \rho) = \lim_{j \to \infty} p(s_j R) = R^{1-n} \int_{B_R} |D\varphi_C|.$$

Thus we have proved that

$$\rho^{1-n} \int_{B_\rho} |D\varphi_C|$$

is independent of ρ and so, from (5.11), we conclude that

$$\int_{\partial B_1} |\varphi_C(\rho x) - \varphi_C(rx)| dH_{n-1} = 0$$

for almost all ρ, $r > 0$. Hence the set C differs only on a set of measure zero from a cone with vertex at the origin.

Note also that from (5.16), (9.4) and (9.5) there exists a constant q such that

$$0 < \omega_{n-1} \leq q \leq \frac{n\omega_n}{2}$$

and

$$\int_{B_\rho} |D\varphi_C| = q\rho^{1-n}$$

and so $0 \in \partial C$. $\square$

The cone C is called a *tangent cone* to E at 0.

If E is regular at 0, then C must be a half space. In fact the converse is also true: If C is a half space then ∂E is regular in a neighbourhood of 0. This will be proved in the next theorem.

9.4 Theorem [MM5]: *Suppose $\{E_j\}$ is a sequence of minimal sets in B_1, converging locally to a minimal set E. Let $x \in \partial^* E$ and let $\{x_j\}$ be a sequence of points such that $x_j \in \partial E_j$ and $x_j \to x$. Then for j sufficiently large x_j is a regular point of ∂E_j and*

$$\lim_{j \to \infty} v^{E_j}(x_j) = v(x).$$

Proof: Let $\sigma(n, \alpha)$ be the constant of Theorem 8.1. As $x \in \partial^* E$, there exists an $R > 0$ such that $B(x, R) \subseteq B_1$ and for every $\rho < R$

$$(9.6) \quad \int_{B(x, \rho)} |D\varphi_E| - | \int_{B(x, \rho)} D\varphi_E | < 2^{-n}\sigma(n, \alpha)\rho^{n-1}$$

(to prove this inequality use 3.3 and (5.14)). On the other hand by Lemma 9.1, we have for almost all $\rho > 0$

$$(9.7) \quad \int_{B(x, \rho)} |D\varphi_E| = \lim_{j \to \infty} \int_{B(x, \rho)} |D\varphi_{E_j}|,$$

$$(9.8) \quad \int_{B(x, \rho)} D\varphi_E = \lim_{j \to \infty} \int_{B(x, \rho)} D\varphi_{E_j}.$$

If we let ρ be such that (9.6), (9.7) and (9.8) all hold, then for j sufficiently large

$$\int_{B(x, \rho)} |D\varphi_{E_j}| - | \int_{B(x, \rho)} D\varphi_{E_j} | < 2^{1-n}\sigma(n, \alpha)\rho^{n-1}.$$

Since x_j converges to x, for j large enough we may assume $|x_j - x| < \rho/2$ and hence $B(x_j, \rho/2) \subseteq B(x, \rho)$, so that

$$\int_{B(x_j, \rho/2)} |D\varphi_{E_j}| - | \int_{B(x_j, \rho/2)} D\varphi_{E_j} | < \sigma(n, \alpha)(\rho/2)^{n-1}.$$

Thus from Theorem 8.4 we deduce the regularity of ∂E_j at x_j.

If $r < \rho/2$ and we denote $v^{E_j}(x)$ by $v^j(x)$, then

$$|v^j(x_j) - v(x)| \le |v_r^j(x_j) - v^j(x_j)| + |v_r^j(x_j) - v_r(x)| + |v_r(x) - v(x)|$$

and so, using Corollary 8.3,

$$(9.9) \quad |v^j(x_j) - v(x)| \le 2\eta(n, \alpha)\sqrt{\frac{2r}{\rho}} + |v_r^j(x_j) - v_r(x)|.$$

Now by (9.7) and (9.8), for almost all r we have

$$\lim_{j \to \infty} v_r^j(x_j) = v_r(x)$$

and hence passing to the limit in (9.9) we see that

$$\limsup_{j \to \infty} |v^j(x_j) - v(x)| \leq 2\eta(n, \alpha) \sqrt{\frac{2r}{\rho}}.$$

The conclusion follows by letting $r \to 0$. $\square$

From the theorems above we immediately have

9.5 Corollary: *Suppose that no singular minimal cones exist in $\mathbb{R}^n$. Then, for every set $E \subseteq \mathbb{R}^n$ with $\psi(E, \rho) = 0$, $\partial E \cap B_\rho$ is an analytic hypersurface.*

We will show that no singular minimal cones can exist in $\mathbb{R}^n$, when $n < 8$, thus proving the regularity of minimal hypersurfaces in $\mathbb{R}^n$, $n \leq 7$. Rather than consider any singular minimal cone we concentrate our attention on those singular minimal cones which only have a singularity at the origin. If the cone C of Theorem 9.3 has other singularities we show that, by again using a "blowing up" procedure, a minimal cylinder can be constructed and from this another minimal cone, singular at the origin but now in $\mathbb{R}^{n-1}$. Continuing this process we eventually get a cone in $\mathbb{R}^{n-k}$ (for some integer k) which is minimal and has just one singular point, namely the origin.

9.6 Proposition: *Suppose C is a minimal cone with vertex at 0 and let $x_0 \in \partial C - \{0\}$. For $t > 0$, let*

$$C_t = \{x \in \mathbb{R}^n : x_0 + t(x - x_0) \in C\}.$$

Then there exists a sequence $\{t_j\}$ converging to zero such that $C_j = C_{t_j}$ converges to a minimal cone Q. Moreover Q is a cylinder with axis through 0 and x_0.

Proof: We may suppose $x_0 = (0, 0, \ldots, 0, a)$, $a \neq 0$. Then we have

$$\varphi_{C_t}(x) = \varphi_C(x_0 + t(x - x_0))$$

and hence

$$\int_{B(x_0, \rho)} |D\varphi_{C_t}| = t^{1-n} \int_{B(x_0, \rho t)} |D\varphi_C| = \rho^{n-1} \int_{B(x_0, 1)} |D\varphi_C|.$$

Then arguing as in Theorem 9.2 we conclude that there exists a sequence $\{t_j\}$

converging to 0 such that C_j converges to a minimal cone Q, with vertex at x_0.

It remains only to prove that Q is a cylinder with axis through 0 and x_0, that is, that there exists a set $A \subseteq \mathbb{R}^{n-1}$ such that $Q = A \times \mathbb{R}$. As C is a cone with vertex at 0, we have $\langle x, D\varphi_C \rangle = 0$ and hence

$$a D_n \varphi_C = -\langle x - x_0, D\varphi_C \rangle.$$

Thus

$$|D_n \varphi_C| \leqq \frac{|x - x_0|}{|x_0|} |D\varphi_C|$$

and therefore

$$\int_{B(x_0,\rho)} |D_n \varphi_{C_t}| = t^{1-n} \int_{B(x_0,\rho t)} |D_n \varphi_C| \leqq \frac{t^{2-n}\rho}{|x_0|} \int_{B(x_0,\rho t)} |D\varphi_C| \leqslant$$

$$\leqq \frac{1}{2} \frac{n\omega_n \rho^n}{|x_0|} t.$$

So we can conclude that

$$D_n \varphi_Q = \lim_{j \to \infty} D_n \varphi_{C_{t_j}} = 0.$$

On the other hand, we have for almost all $s < t$, by the proof of Lemma 2.4,

$$\int_{\mathcal{B}_R} |\varphi_s - \varphi_t| dH_{n-1} \leqq \int_{\mathcal{B}_R \times (s,t)} |D_n \varphi_Q| = 0$$

where $\varphi_r(y) = \varphi_Q(y, r)$. Hence there exists a set $A \subseteq \mathbb{R}^{n-1}$ such that for almost all r and s we have

$$\varphi_Q(y, s) = \varphi_Q(y, r) = \varphi_A(y)$$

for almost all $y \in \mathbb{R}^{n-1}$. Thus

$$Q = A \times \mathbb{R}. \qquad \qquad \qquad \square$$

9.7   Remark: Since Q itself is a cone, we have for each $t > 0$, $(y, s) \in \mathbb{R}^{n-1} \times \mathbb{R}$

$$\varphi_A(ty) = \varphi_Q(ty, ts) = \varphi_Q(y, s) = \varphi_A(y)$$

and so A is also a cone. We show that A is minimal if and only if Q is minimal. To do this we need the following

9.8   Lemma: *Suppose* $f \in BV_{loc}(\mathbb{R}^n)$ *and let* Ω *be a bounded open set in* $\mathbb{R}^{n-1}$. *Then we have*

$$(9.10) \qquad \int_{\Omega \times (-T, T)} |Df| \geq \int_{-T}^{T} dt \int_{\Omega} |Df_t|$$

where $f_t(y) = f(y, t)$, *with equality holding if* f *is independent of* x_n.

Proof: If $\Lambda = \Omega \times (-T, T)$ and $f \in C^1(\Lambda)$, then inequality (9.10) is trivial. Now suppose $f \in BV(\Lambda)$ and let $\{f_j\}$ be a sequence of C^1-functions such that $f_j \to f$ in $L_1(\Lambda)$ and

$$\lim_{j \to \infty} \int_{\Lambda} |Df_j| dx = \int_{\Lambda} |Df| \qquad \text{(Theorem 1.17)}.$$

Passing, possibly, to a subsequence, we can assume that for almost all t we have

$$f_{j,t} \to f_t \quad \text{in} \quad L_1(\Omega).$$

Therefore, by Theorem 1.9,

$$\int_{\Omega} |Df_t| \leq \liminf_{j \to \infty} \int_{\Omega} |Df_{j,t}| dx$$

and (9.10) is proved.

If f is independent of x_n, then we have

$$\int f D_n g \, dx = 0 \qquad \text{for} \quad g \in C_0^1(\Lambda)$$

and, if $|g| \leq 1$ and $g \in C_0^1(\Lambda; \mathbb{R}^n)$.

$$\int f \operatorname{div} g \, dx = \int_{-T}^{T} dt \int_{\Omega} f_t \operatorname{div} g_t dy \leq \int_{-T}^{T} dt \int_{\Omega} |Df_t|.$$

Hence equality must hold in (9.10).   □

9.9 Proposition: *Suppose $Q = A \times \mathbb{R}$ is a cylinder in $\mathbb{R}^n = \mathbb{R}^{n-1} \times \mathbb{R}$. Then Q is minimal in $\mathbb{R}^n$ if and only if A is minimal in $\mathbb{R}^{n-1}$.*

Proof:
(I) Suppose A is minimal in $\mathbb{R}^{n-1}$ and let M be a Caccioppoli set in $\mathbb{R}^n$ coinciding with Q outside some compact set K. Choose T such that

$$K \subseteq \Lambda = \mathscr{B}_T \times (-T, T).$$

Now from Lemma 9.8

$$\int_\Lambda |D\varphi_M| \leq \int_{-T}^{T} dt \int_{\mathscr{B}_T} |D\varphi_{M_t}|$$

where $M_t \subseteq \mathbb{R}^{n-1}$ is defined by

$$\varphi_{M_t}(y) = \varphi_M(y, t).$$

We have $M_t = A$ outside a compact set $H_t \subseteq \mathscr{B}_T$ and hence

$$\int_{\mathscr{B}_T} |D\varphi_A| \leq \int_{\mathscr{B}_T} |D\varphi_{M_t}|.$$

Thus in conclusion

$$\int_\Lambda |D\varphi_M| \geq \int_{-T}^{T} dt \int_{\mathscr{B}_T} |D\varphi_A| = \int_\Lambda |D\varphi_Q|$$

so that Q is minimal.

(II) Suppose now that Q is minimal. If A is not minimal in $\mathbb{R}^{n-1}$, there exist $\varepsilon > 0$, $R > 0$ and a set E coinciding with A outside some compact set $H \subseteq \mathscr{B}_R$ such that

$$\int_{\mathscr{B}_R} |D\varphi_E| \leqq \int_{\mathscr{B}_R} |D\varphi_A| - \varepsilon.$$

Let T be any number greater than zero and set

$$M = \begin{cases} E \times (-T, T) & \text{in } |x_n| < T \\ Q & \text{outside } |x_n| < T \end{cases}$$

so that we have $M = Q$ outside $H \times [-T, T]$. Hence

(9.11) $$\int_{\mathscr{B}_R \times [-T, T]} |D\varphi_Q| \leqq \int_{\mathscr{B}_R \times [-T, T]} |D\varphi_M|.$$

On the other hand, we have

$$\int_{\mathscr{B}_R \times [-T, T]} |D\varphi_Q| = 2T \int_{\mathscr{B}_R} |D\varphi_A|$$

and

(9.12) $$\int_{\mathscr{B}_R \times [-T, T]} |D\varphi_M| \leq 2T \int_{\mathscr{B}_R} |D\varphi_E| + 2\omega_{n-1} R^{n-1}$$

$$\leqq 2T \int_{\mathscr{B}_R} |D\varphi_A| - 2T\varepsilon + 2\omega_{n-1} R^{n-1}$$

which contradicts (9.11) for sufficiently large T.   □

Suppose now that C is a minimal cone in $\mathbb{R}^n$, singular at 0 and also at the point $x_0 \neq 0$. As C is a cone, all the points on the half line through 0 and x_0 must be singular. By a rotation we may suppose that this half line is the positive x_n axis. Now, if we blow up C near x_0, we obtain a minimal cylinder Q such that the x_n axis lies on ∂Q and all the points on the x_n axis are singular (as they are the limit of singular points-Theorem 9.4). From Propositions 9.6 and 9.9 it follows that we may write $Q = A \times \mathbb{R}$, where A is a minimal cone in R^{n-1}, singular at the origin. Repeating the argument above if necessary we have the following

9.10   Theorem: *Suppose C is a minimal cone in $\mathbb{R}^n$, singular at 0. Then there exists $k \leqq n$ and a cone $A \subseteq \mathbb{R}^k$ such that A is minimal and has just one singularity, namely 0.*

In the next chapter we shall prove that such a cone cannot exist if $k \leqq 7$ and thus, by Corollary 9.5 and Theorem 9.10, prove the regularity of minimal surfaces in $\mathbb{R}^n$, $n \leqq 7$.

10. The First and Second Variation of the Area

Suppose $E \subseteq \mathbb{R}^n$ is a minimal set in B_1, and $\{F_t\}$ is a one parameter family of diffeomorphisms $\mathbb{R}^n \to \mathbb{R}^n$ such that $F_0 = I = $ identity and the maps $F_t - I$ have uniform compact support in B_1. The sets

$$E_t = F_t(E) = \{F_t(x) : x \in E\}$$

must equal E outside B_1 and so

$$\int_{B_1} |D\varphi_E| \leq \int_{B_1} |D\varphi_{E_t}|$$

Then, assuming appropriate smoothness, we see that the function

$$A(t) = \int_{B_1} |D\varphi_{E_t}|$$

has a minimum at zero and so

(10.1) $\quad \dfrac{d}{dt} A(t)|_{t=0} = 0$

and

(10.2) $\quad \dfrac{d^2}{dt^2} A(t)|_{t=0} \leq 0.$

In this chapter we shall consider the case where E is a cone in $\mathbb{R}^n$, smooth everywhere except possibly at 0. Then, by constructing a particular diffeomorphism and using (10.1) and (10.2), we show that either ∂E is a hyperplane or $n \geq 8$. This important theorem was first proved by Simons in 1968 [SJ].

Firstly we calculate the first and second variations of the area for general sets and then later consider the special case of the cone.

10.1 Lemma: *Suppose $\Omega \subseteq \mathbb{R}^n$, f is a function in $BV_{\text{loc}}(\Omega)$ and $F : \mathbb{R}^n \to \mathbb{R}^n$ is a diffeomorphism. Suppose $A \subset\subset \Omega$ and let $f_* = f \circ F^{-1}$, $\Omega_* = F(\Omega)$ and $A_* = F(A)$. Then*

$$(10.3) \quad \int_{A_*} |Df_*| = \int_A |HDf|$$

where $H = |\det DF|[DF]^{-1}$.

Proof: Let $f \in C^1(\Omega)$ and $g \in C_0^1(A; \mathbb{R}^n)$. Then, setting $\phi = F^{-1}$,

$$(10.4) \quad \int < g_*, Df_* > dx = \int < g \circ \phi, D\phi \; Df \circ \phi > dx$$

$$= \int < g, (D\phi \circ F) \; Df > |\det DF| dy = \int < g, H \; Df > dy.$$

Thus we see that (10.3) holds for functions f in $C^1(\Omega)$ and indeed it is just the change of area formula for such functions.

Now suppose only that $f \in BV_{\text{loc}}(\Omega)$. We can approximate f by a sequence $\{f_j\}$ of C^1 functions (see Theorem 1.17) which implies that the corresponding functions f_{j_*} converge to f_* in $L^1(A_*)$. Hence, passing to the limit in (10.4),

$$\int < g_*, Df_* > \; = \int < g, HDf > \; = \int < g, Hv > |Df|,$$

where v is obtained by differentiating Df with respect to $|Df|$ (and so $|v| = 1$ $|Df|$-almost everywhere).

Now suppose that $|g| \leq 1$ and $g \in C_0^1(A; \mathbb{R}^n)$. Then $|g_*| \leq 1$ and spt $g_* \subseteq A_*$, so that

$$\int < g, Hv > |Df| \leq \int_{A_*} |Df_*|.$$

Taking the supremum of the left hand side gives

$$\int_A |HDf| = \int_A |Hv| \; |Df| \leq \int_{A_*} |Df_*|.$$

To show the reverse inequality, we notice that, if $\gamma \in C_0^1(A_*; \mathbb{R}^n)$, $|\gamma| \leq 1$ and we put $g = \gamma \circ F \in C_0^1(A; \mathbb{R}^n)$, then $g_* = \gamma$ and

$$\int < \gamma, Df_* > \; \leq \int_A |Hv| \; |Df|.$$

Hence

$$\int_{A_*} |Df_*| \le \int_A |Hv|\,|Df|.$$ □

10.2: We can use this change of area formula to obtain expressions for the first and second variation as follows:

Let F_t be a one-parameter family of diffeomorphisms of $\mathbb{R}^n$ such that $F_0 = I$ and put $\phi_t = F_t^{-1}$. Suppose also that there exists a fixed compact set K such that $F_t = I$ outside K for $t \in [0, 1]$. If $A \supseteq K$ then we must have that $A_{*t} = F_t(A) = A$ and, from Lemma 10.1, we easily obtain

$$(10.5) \quad \left\{ \frac{d}{dt} \int_A |Df_{*t}| \right\}_{t=0} = \int_A <v, \dot{H}_0 v> |Df|,$$

$$(10.6) \quad \left\{ \frac{d^2}{dt^2} \int_A |Df_{*t}| \right\}_{t=0} = \int_A \left\{ |\dot{H}_0 v|^2 + <v, \ddot{H}_0 v> - <v, \dot{H}_0 v>^2 \right\} |Df|$$

where

$$\dot{H}_0 = \left(\frac{dH}{dt} \right)_{t=0}, \quad \ddot{H}_0 = \left(\frac{d^2 H}{dt^2} \right)_{t=0}$$

In particular, if $f = \varphi_E$ and we define the set E_t by $\varphi_{E_t} = (\varphi_E)_{*t}$, we have $E_t = E$ outside $K \subseteq A$ and

$$(10.7) \quad \left\{ \frac{d}{dt} \int_A |D\varphi_{E_t}| \right\}_{t=0} = \int_{A \cap \partial^* E} <v, \dot{H}_0 v> dH_{n-1},$$

$$(10.8) \quad \left\{ \frac{d^2}{dt^2} \int_A |D\varphi_{E_t}| \right\}_{t=0} = \int_{A \cap \partial^* E} \left\{ |\dot{H}_0 v|^2 + <v, \ddot{H}_0 v> - <v, \dot{H}_0 v>^2 \right\} dH_{n-1}.$$

10.3 Notation: For the theorems and calculations to follow it is convenient to introduce the *tangential derivatives*:

For $x \in \partial^* E$ define

$$\delta_i = D_i - v_i \sum_{h=1}^{n} v_h D_h$$

where $D_h = \dfrac{\partial}{\partial x_h}$ and v is the normal vector at x.

If we have a function $g:\mathbb{R}^n \to \mathbb{R}$, then the vector $\delta g = (\delta_1 g, \delta_2 g, \ldots, \delta_n g)$ may be written

$$\delta g = Dg - v\langle Dg \cdot v\rangle$$

so that δg is Dg minus the component of Dg in the normal direction. Thus δg, as its name indicates, is merely the component of Dg in the tangential direction.

We remark that δg depends only on the values of g on $\partial^* E$, or in other words if $g = 0$ on $\partial^* E$ then $\delta g = 0$. To see that, let $x_0 \in \partial^* E$. If $Dg(x_0) = 0$ then $\delta g(x_0) = 0$; otherwise $\partial^* E$ is a smooth hypersurface near x_0 with normal vector

$$v(x) = Dg(x)/|Dg(x)|$$

and again $\delta g(x_0) = 0$.

It is also convenient to employ the summation convention, that is that in any expression repeated indices indicate summation from 1 to n. With this notation we may write

$$\delta_i = D_i - v_i v_h D_h.$$

From (10.7) and (10.8) using the notation described above we obtain the following result.

10.4 Theorem: *Suppose $F_t = I + tg$ where $g:\mathbb{R}^n \to \mathbb{R}^n$ has compact support. Then, if $A \supseteq spt\, g$,*

$$(10.9) \quad \left\{\frac{d}{dt}\int_A |D\varphi_{E_t}|\right\}_{t=0} = \int_{\partial^* E \cap A} \delta_i g_i dH_{n-1},$$

$$(10.10) \quad \left\{\frac{d^2}{dt^2}\int_A |D\varphi_{E_t}|\right\}_{t=0} = \int_{\partial^* E \cap A} \left\{(\delta_i g_i)^2 - \delta_i g_j \delta_j g_i + v_i v_h \delta_j g_i \delta_j g_h\right\}dH_{n-1}.$$

(Note that δ_i applies only to the symbol immediately following it.)

Proof: We have $DF_t = I + tDg$ and hence

$$(DF_t)^{-1} = I - tDg + t^2 DgDg + 0(t^3)$$

and

$$\det DF_t = 1 + t\tau(Dg) + \frac{1}{2}t^2[\tau(Dg)^2 - \tau(DgDg)] + 0(t^3),$$

where

$$\tau(A) = \text{trace } (A) = \sum_{i=1}^{n} A_{ii} = A_{ii}.$$

Therefore

$$\dot{H}_0 = \tau(Dg)I - Dg,$$

$$\ddot{H}_0 = 2DgDg - 2\tau(Dg)Dg + [\tau(Dg)^2 - \tau(DgDg)]I,$$

so that we have immediately

$$\langle v, \dot{H}_0 v \rangle = \tau(Dg) - v_i v_j D_i g_j = \delta_i g_i$$

and (10.9) is proved.

To prove (10.10) we put div $g = D_i g_i$ to obtain

$$|\dot{H}_0 v|^2 + \langle v, \ddot{H}_0 v \rangle = 2(\text{div } g)^2 - 4(\text{div } g)v_i v_h D_i g_h +$$

$$+ v_i v_h D_j g_h D_j g_i + 2v_i v_h D_j g_h D_i g_j - D_i g_j D_j g_i =$$

$$= 2(\delta_i g_i)^2 - 2v_i v_j v_h v_k D_i g_j D_h g_k + v_i v_h \delta_j g_i \delta_j g_h +$$

$$+ 2v_i v_j v_h v_k D_i g_j D_h g_k - \delta_i g_j \delta_j g_i =$$

$$= 2(\delta_i g_i)^2 - \delta_i g_j \delta_j g_i + v_i v_h \delta_j g_i \delta_j g_h$$

and (10.10) now follows. □

We now choose a particular deformation and obtain the corresponding variation formulas. The deformation we choose will be one which shifts the original set in a direction normal to the surface.

10.5: Suppose ζ is a function in $C_0^1(A)$ such that spt $\zeta \cap \partial E$ is a C^2-hypersurface. Then, if we set

$$d(x) = \begin{cases} \text{dist } (x, \partial E) & x \in E \\ -\text{dist } (x, \partial E) & x \in \mathbb{R}^n - E, \end{cases}$$

d is C^2 in a neighbourhood of spt $\zeta \cap \partial E$ (see Appendix B, $[GT]$) and we have

$$v = Dd \qquad \text{in spt } \zeta \cap \partial E.$$

Now since $1 = |v|^2 = d_h d_h$ in spt $\zeta \cap \partial E$ we obtain that

$$d_h d_{ih} = \frac{1}{2} D_i |v|^2 = 0.$$

Hence

$$\delta_i v_j = d_{ij} - d_i d_h d_{jh} = d_{ij}$$

and we see that

(10.11) $\delta_i v_j = \delta_j v_i.$

Choosing $g = \zeta v$ in (10.9) and (10.10) gives

$$(10.12) \quad \left\{ \frac{d}{dt} \int_A |D\varphi_{E_t}| \right\}_{t=0} = \int_{\partial E} \mathscr{H} \zeta dH_{n-1},$$

where $\mathscr{H} = \delta_i v_i$, and

$$(10.13) \quad \left\{ \frac{d^2}{dt^2} \int_A |D\varphi_{E_t}| \right\}_{t=0} = \int_{\delta E} \left\{ |\delta \zeta|^2 - (c^2 - \mathscr{H}^2)\zeta^2 \right\} dH_{n-1}$$

where $c^2 = (\delta_i v_j)(\delta_j v_i) = \sum_{i,j} (\delta_i v_j)^2.$

The last two equations follow from the identities

(10.14) $v_i \delta_i = 0$

(10.15) $v_i \delta_j v_i = 0 \qquad j = 1, \ldots, n.$

10.6   Remark: It can be shown that $\mathscr{H} = \mathscr{H}(x)$ is just the mean curvature of the surface ∂E at x, and $c^2 = c^2(x)$ is the sum of the squares of the principal curvatures of ∂E calculated at x.

The identities in the next lemma will also be useful in what follows.

10.7   Lemma: *Suppose ∂E is a smooth hypersurface in a neighbourhood of x_0. Then at x_0 we have*

(10.16) $\delta_i\delta_j = \delta_j\delta_i + (v_i\delta_j v_h - v_j\delta_i v_h)\delta_h$

(10.17) $\mathscr{D}v_j \equiv \delta_i\delta_i v_j = -c^2 v_j + \delta_j\mathscr{H}.$

Proof: Substituting for δ_j we obtain

$$\delta_i\delta_j = \delta_i D_j - v_j v_h \delta_i D_h - (v_h\delta_i v_j + v_j\delta_i v_h)D_h =$$
$$= (D_i D_j - v_i v_h D_h D_j) - (v_j v_h D_i D_h - v_j v_h v_i v_k D_k D_h) -$$
$$- (v_h\delta_i v_j + v_j\delta_i v_h)D_h.$$

Noting that $\delta_i v_j = \delta_j v_i$ and of course $D_i D_j = D_j D_i$ and cancelling symmetric terms (in i and j), we have

$$\delta_i\delta_j - \delta_j\delta_i = v_i\delta_j v_h D_h - v_j\delta_i v_h D_h = (v_i\delta_j v_h - v_j\delta_i v_h)D_h.$$

To obtain the equation with δ_h replacing D_h we merely notice that $v_h\delta_i v_h = 0$ (see (10.15)) for $i = 1, \ldots, n$, and (10.16) is proved.

Now considering (10.17) and using (10.11), (10.16) and (10.14), we have

$$\delta_i\delta_i v_j = \delta_i\delta_j v_i = \delta_j\delta_i v_i + (v_i\delta_j v_h - v_j\delta_i v_h)\delta_h v_i = \delta_j\mathscr{H} - v_j c^2. \qquad \square$$

10.8   Lemma (*Integration by parts*)*: Let ∂E be a smooth hypersurface and let $\varphi \in C_0^1(\mathbb{R}^n)$. Then*

$$\int_{\partial E} \delta_i\varphi \, dH_{n-1} = - \int_{\partial E} \varphi v_i dH_{n-1}.$$

Proof: Suppose first that $\partial E \cap \operatorname{spt}\varphi$ is the graph of a function $x_n = g(\bar x)$,

$\bar{x} = (x_1, x_2, \ldots, x_{n-1}) \in U$. We have $v_n = (1 + |Dg|^2)^{-1/2}$; $v_i = -v_n D_i g$ and dH_{n-1}
$= v_n^{-1} d\bar{x}$, so that

$$\int_{\partial E} \delta_i \varphi dH_{n-1} = \int_U (\delta_{ih} - v_i v_h) v_n^{-1} D_h \varphi d\bar{x} = -\int_U \varphi D_h [v_n^{-1}(\delta_{ih} - v_i v_h)] d\bar{x} =$$

$$= -\int_U \varphi v_n^{-1} v_i D_h v_h d\bar{x} = -\int_{\partial E} \mathscr{H} v_i \varphi dH_{n-1}$$

since $D_i v_n + v_h v_n^2 D_h(v_i/v_n) = 0$.

The general case follows easily by a partition of unity. $\square$

In particular, if ∂E is stationary (that is the first variation vanishes or $\mathscr{H} = 0$)
we have the standard formula of integration by parts:

$$\int_{\partial E} u \delta_i v dH_{n-1} = -\int_{\partial E} v \delta_i u dH_{n-1}$$

provided uv has compact support. Moreover, since $v_i \delta_i = 0$, we can always
integrate by parts with the Laplace operator $\mathscr{D}$ to obtain:

$$\int_{\partial E} u \mathscr{D} v dH_{n-1} = -\int_{\partial E} \delta_i u \delta_i v dH_{n-1} = \int_{\partial E} v \mathscr{D} u dH_{n-1}$$

whenever uv has compact support.

When ∂E is stationary we obtain from (10.16) by means of simple calculations
the identity:

$$\mathscr{D} \delta_h = \delta_h \mathscr{D} - 2v_h(\delta_i v_j)\delta_i \delta_j - 2(\delta_h v_j)(\delta_j v_i)\delta_i.$$

Moreover in this case the second variation (10.13) is given by

$$\int_{\partial E} \{|\delta \zeta|^2 - c^2 \zeta^2\} dH_{n-1}.$$

We want to examine this expression more closely in the particular case where
E is a cone. To do this, it is necessary to estimate the term $\mathscr{D} c^2$ which we do
in the next lemma.

10.9   Lemma: *Suppose E is a cone which is stationary and suppose A is an open
set such that $\partial E \cap A$ is regular. Then in A we have*

$$(10.19) \quad \frac{1}{2}\mathscr{D}c^2 \geqq -c^4 + |\delta c|^2 + \frac{2c^2}{|x|^2}.$$

Proof: By the definition of c^2

$$\frac{1}{2}\mathscr{D}c^2 = (\delta_i v_j)\mathscr{D}\delta_i v_j + (\delta_h \delta_i v_j)(\delta_h \delta_i v_j).$$

Then, using (10.18), (10.17) and the fact $\mathscr{H} = 0$, we have

$$\frac{1}{2}\mathscr{D}c^2 = -(\delta_i v_j)\delta_i(c^2 v_j) - 2(\delta_i v_j)(\delta_k v_h)(\delta_h v_j)(\delta_i v_k) + (\delta_h \delta_i v_j)(\delta_h \delta_i v_j)$$

and by (10.16)

$$= -c^4 - 2v_i v_h(\delta_j \delta_h v_k)(\delta_k \delta_i v_j) + (\delta_h \delta_i v_j)(\delta_h \delta_i v_j).$$

Now, if $x_0 \in \partial E \cap A$, we can choose the x_n-axis to be the same direction as $v(x_0)$. At x_0, we then have

$$v_n = 1, \ \delta_n = 0, \ v_\alpha = 0, \ \delta_\alpha = D_\alpha \qquad \alpha = 1, \dots, n-1$$

and, using (10.16) and (10.15),

$$\delta_i \delta_n v_n = 0.$$

Hence at x_0

$$\frac{1}{2}\mathscr{D}c^2 = -c^4 + (\delta_\gamma \delta_\alpha v_\beta)(\delta_\gamma \delta_\alpha v_\beta) + 2(\delta_\gamma \delta_\alpha v_n)(\delta_\gamma \delta_\alpha v_n) - 2(\delta_\alpha \delta_\beta v_n)(\delta_\alpha \delta_\beta v_n)$$

$$= -c^4 + (\delta_\gamma \delta_\alpha v_\beta)(\delta_\gamma \delta_\alpha v_\beta)$$

where all the greek indices indicate summation from 1 to $n-1$.

On the other hand we have

$$|\delta c|^2 = \frac{(\delta_\alpha v_\beta)(\delta_\gamma \delta_\alpha v_\beta)(\delta_\sigma v_\tau)(\delta_\gamma \delta_\sigma v_\tau)}{c^2}$$

and hence

$$\frac{1}{2}\mathscr{D}c^2 + c^4 - |\delta c|^2 =$$

$$= \frac{[(\delta_\sigma v_\tau)(\delta_\gamma \delta_\alpha v_\beta) - (\delta_\alpha v_\beta)(\delta_\gamma \delta_\sigma v_\tau)][(\delta_\sigma v_\tau)(\delta_\gamma \delta_\alpha v_\beta) - (\delta_\alpha v_\beta)(\delta_\gamma \delta_\sigma v_\tau)]}{2c^2}$$

Now suppose that E is a cone with vertex at the origin. We can choose coordinates in such a way that x_0 lies on the $(n-1)$-axis. As E is a cone, $\langle x, v \rangle = 0$ on ∂E. Hence $v_{n-1} = 0$ at x_0 (which we had already assumed above) and

$$0 = \delta_i \langle x, v \rangle = \langle \delta_i x, v \rangle + \langle x, \delta_i v \rangle = \langle x, \delta_i v \rangle$$

so that

$$\delta_i v_{n-1} = 0 \qquad \text{at} \quad x_0.$$

If the letters A, B, S, T run from 1 to $n-2$, we have

$$\frac{1}{2}\mathscr{D}c^2 + c^4 - |\delta c|^2 =$$

$$= \frac{[(\delta_S v_T)(\delta_\gamma \delta_A v_B) - (\delta_A v_B)(\delta_\gamma \delta_S v_T)][(\delta_S v_T)(\delta_\gamma \delta_A v_B - (\delta_A v_B)(\delta_\gamma \delta_S v_T)]}{2c^2} +$$

$$+ \frac{2}{c^2}(\delta_S v_T)(\delta_S v_T)(\delta_\gamma \delta_{n-1} v_\alpha)(\delta_\gamma \delta_{n-1} v_\alpha) \geqq 2(\delta_\gamma \delta_{n-1} v_\alpha)(\delta_\gamma \delta_{n-1} v_\alpha).$$

From (10.16) and the assumptions above,

$$\delta_i \delta_{n-1} = \delta_{n-1} \delta_i$$

and, as x_0 lies on the x_{n-1}-axis,

$$\delta_{n-1} = D_{n-1} = \pm \frac{x_i D_i}{|x|} \qquad \text{at} \quad x_0.$$

Since E is a cone, the function v is homogeneous of degree 0 and hence $\delta_i v_\alpha$ is homogeneous of degree -1. Thus it follows by Euler's theorem on homogeneous functions that

$$\delta_{n-1}\delta_i v_\alpha = \pm \frac{x_j D_j}{|x|}\delta_i v_\alpha = \mp \frac{1}{|x|}\delta_i v_\alpha$$

and hence

$$(\delta_i \delta_{n-1} v_\alpha)(\delta_i \delta_{n-1} v_\alpha) = \frac{1}{|x|^2}(\delta_i v_\alpha)(\delta_i v_\alpha) = \frac{c^2}{|x|^2}$$

proving (10.19). $\square$

We are now able to prove the following important theorem of Simons.

10.10   Theorem: (Simons [SJ]): *Suppose E is a cone, such that ∂E is regular in $\mathbb{R}^n - \{0\}$. Suppose that $\mathscr{H} \equiv 0$ and that the second variation of the area is non-negative, that is*

$$(10.20)\quad \int_{\partial E} \{|\delta\zeta|^2 - c^2\zeta^2\}dH_{n-1} \geq 0$$

for every $\zeta \in C_0^1$, with spt $\zeta \cap \{0\} = \varnothing$. Then either ∂E is a hyperplane or $n \geq 8$.

Proof: Replacing ζ by ζc in (10.20), we have

$$\int_{\partial E} c^4 \zeta^2 dH_{n-1} \leq \int_{\partial E} \left\{ c^2|\delta\zeta|^2 + \zeta^2|\delta c|^2 + \frac{1}{2} < \delta c^2, \delta\zeta^2 > \right\}dH_{n-1}$$

$$= \int_{\partial E} \left\{ c^2|\delta\zeta|^2 + \zeta^2|\delta c|^2 - \frac{1}{2}\zeta^2 \mathscr{D}c^2 \right\}dH_{n-1},$$

where at the last step we have used integration by parts. Now using (10.19) we obtain

$$(10.21)\quad 0 \leq \int_{\partial E} \left\{ |\delta\zeta|^2 - \frac{2\zeta^2}{|x|^2} \right\}c^2 dH_{n-1}$$

for every $\zeta \in C_0^1$ with spt $\zeta \cap \{0\} = \varnothing$. Obviously (10.21) will hold by approximation for every function ζ provided that

$$\int_{\partial E} \frac{\zeta^2}{|x|^2} c^2 dH_{n-1} < \infty.$$

In the previous lemma we showed that $\delta_i \nu_\alpha$ was homogeneous of degree -1 and so c^2 is homogeneous of degree -2. Thus (10.21) holds for any ζ such that

$$\int_{\partial E} \zeta^2 |x|^{-4} dH_{n-1} < \infty.$$

If $r_1 = \max\{|x|, 1\}$, $r = |x|$ and setting

$$\zeta = r^\alpha r_1^\beta,$$

then

$$\int_{\partial E} \zeta^2 |x|^{-4} dH_{n-1} = H_{n-2}(\partial E \cap \partial B_1) \int_0^\infty r^{2\alpha} r_1^{2\beta} r^{n-6} dr.$$

Thus to make the integral finite, we must have

$$2\alpha + n - 6 > -1.$$

$$2(\alpha + \beta) + n - 6 < -1$$

and hence

$$(10.22) \quad \begin{cases} \alpha > \dfrac{5-n}{2} \\[2mm] \alpha + \beta < \dfrac{5-n}{2} \end{cases}$$

If inequalities (10.22) hold then (10.21) becomes

$$0 \leq (\alpha^2 - 2) \int_{\partial E \cap B_1} r^{2\alpha - 2} c^2 dH_{n-1} + [(\alpha + \beta)^2 - 2] \int_{\partial E - B_1} r^{2(\alpha + \beta) - 2} c^2 dH_{n-1}.$$

If $(\frac{5-n}{2})^2 < 2$, we can always choose α and β satisfying (10.22) but such that $\alpha^2 < 2$ and $(\alpha + \beta)^2 < 2$. It then follows that either $c^2 = 0$, and so ∂E is a hyperplane, or $(5-n)^2 > 8$, that is $n \geq 8$.   □

From this result and Theorem 9.10, it follows that no singular minimal cones can exist in $\mathbb{R}^n$, for $n \leq 7$, as any such minimal cone must satisfy $\mathscr{H} = 0$ and (10.20). Thus we have proved

10.11   Theorem: *Suppose $n \leq 7$ and E is a minimal set in B_ρ (that is $\psi(E, \rho) = 0$). Then $\partial E \cap B_\rho$ is an analytic hypersurface.*

11. The Dimension of the Singular Set

In the last chapter we proved a regularity theorem for dimensions $n \leq 7$. However we said nothing about the regularity in higher dimensions which is our task in this chapter. We prove that the H_k measure of the singular set is zero for all $k > n - 8$ and so actually include the results of the previous chapter.

Some results and definitions concerning Hausdorff measures are needed to prove our main theorem.

11.1 Definition: Suppose $A \subseteq \mathbb{R}^n$, $0 \leq k < \infty$ and $0 < \delta \leq \infty$. Then we define

$$(11.1) \quad H_k^\delta(A) = \omega_k 2^{-k} \inf \left\{ \sum_{j=1}^\infty (\text{diam } S_j)^k : A \subseteq \sum_{j=1}^\infty S_j, \text{ diam } S_j < \delta \right\}$$

and

$$(11.2) \quad H_k(A) = \lim_{\delta \to 0} H_k^\delta(A) = \sup_\delta H_k^\delta(A)$$

where $\omega_k = \Gamma(\tfrac{1}{2})^k / \Gamma(\tfrac{k}{2} + 1)$, $k \geq 0$, is the measure of the unit ball in $\mathbb{R}^n$. H_k is of course the k-dimensional Hausdorff measure.

It is obvious from the definition that $H_k^\rho \geq H_k^\delta$ if $\delta > \rho$, but in fact we can show

11.2 Lemma: *For every* $A \subseteq \mathbb{R}^n$, $H_k^\infty(A) = 0$ *if and only if* $H_k(A) = 0$.

Proof: By the remark above we need only prove that $H_k^\infty(A) = 0$ implies $H_k(A) = 0$.

Let $\varepsilon > 0$. Then there exists a countable covering $\{S_j\}$ of A such that

$$\sum_j (\text{diam } S_j)^k < \varepsilon^k.$$

Clearly then we must have diam $S_j < \varepsilon$ for every j and hence

$$H_k^\varepsilon(A) \leq \omega_k 2^{-k} \varepsilon^k.$$

Letting $\varepsilon \to 0$, the result follows.   □

11.3   Proposition: *If* $A \subseteq \mathbb{R}^n$, *then for* H_k-*almost all* $x \in A$ *we have*

(11.3) $\displaystyle \limsup_{r \to 0} \frac{H_k^{\infty}(A \cap B(x, r))}{\omega_k r^k} \geqq 2^{-k}.$

Proof [FH2]: Define

$$\zeta(S) = \omega_k 2^{-k} (\text{diam } S)^k$$

and, for $t > 0$, $\varepsilon > 0$,

$$B(H_k^{\delta}, t, \varepsilon) = \{x \in A : H_k^{\delta}(S \cap A) \leq t\zeta(S), \text{ for all sets } S$$
$$\text{such that } x \in S \text{ and diam } S < \varepsilon\}.$$

If S is any set in A, with diam $S < \varepsilon$ we have

$$H_k^{\delta}(S \cap B(H_k^{\delta}, t, \varepsilon)) \leq t\zeta(S)$$

and therefore from the definition of H_k^{ε}:

(11.4) $H_k^{\delta}(B(H_k^{\delta}, t, \varepsilon)) \leqq tH_k^{\varepsilon}(B(H_k^{\delta}, t, \varepsilon))$

and so, in particular, if $\delta < 1$

$$H_k^{\delta}(B(H_k^{\delta}, 1 - \delta, \delta)) \leqq (1 - \delta)H_k^{\delta}(B(H_k^{\delta}, 1 - \delta, \delta)).$$

Hence

$$H_k^{\delta}(B(H_k^{\delta}, 1 - \delta, \delta)) = 0$$

and, by Lemma 11.2,

(11.5) $H_k(B(H_k^{\delta}, 1 - \delta, \delta)) = 0.$

Now consider the set

$$C = \left\{ x \in A : \inf_{\varepsilon > 0} \sup \left\{ \frac{H_k^\infty(A \cap S)}{\zeta(S)} : x \in S, \text{ diam } S < \varepsilon \right\} < 1 \right\} =$$

$$= \left\{ x \in A : \inf_{n \in \mathbb{N}} \sup \left\{ \frac{H_k^\infty(A \cap S)}{\zeta(S)} : x \in S, \text{ diam } S < \frac{1}{n} \right\} < 1 \right\}.$$

If $x \in C$ we must have that for some n

$$\sup \left\{ \frac{H_k^\infty(A \cap S)}{\zeta(S)} : x \in S, \text{ diam } S < \frac{1}{n} \right\} < 1 - \frac{1}{n}$$

and hence $x \in B(H_k^\infty, 1 - \frac{1}{n}, \frac{1}{n})$. In conclusion

$$C = \bigcup_{n=1}^{\infty} B\left(H_k^\infty, 1 - \frac{1}{n}, \frac{1}{n} \right) \subseteq \bigcup_{n=1}^{\infty} B\left(H_k^{1/n}, 1 - \frac{1}{n}, \frac{1}{n} \right)$$

and so $H_k(C) = 0$.

If we fix S such that $d = \text{diam } S$ and suppose $x \in S$, then obviously

$$\frac{H_k^\infty(A \cap S)}{\zeta(S)} \leqq \frac{H_k^\infty(A \cap B(x, d))}{\omega_k d^k} 2^k$$

and hence

$$\sup \left\{ \frac{H_k^\infty(A \cap S)}{\zeta(S)} : x \in S, \text{ diam } S < \delta \right\} \leqslant$$

$$\leqq 2^k \sup \left\{ \frac{H_k^\infty(A \cap B(x, \rho))}{\omega_k \rho^k} : \rho < \delta \right\}.$$

Thus the set

$$D = \left\{ x \in A : \limsup_{r \to 0} \frac{H_k^\infty(A \cap B(x, r))}{\omega_k r^k} < 2^{-k} \right\}$$

must be contained in C. Hence $H_k(D) = 0$ and the lemma is proved. $\qquad\square$

We can now start on the study of minimal surfaces. Firstly we show that, if a sequence of minimal sets $\{E_j\}$ converges locally to a minimal set E, then locally the set of singular points of E_j must lie close to the singular points of E. More precisely

11.4   Lemma: *Suppose $\{E_j\}$ is a sequence of minimal sets in Ω converging in $L^1_{\text{loc}}(\Omega)$ to a minimal set E. Let $\Sigma_j = \partial E_j - \partial^* E_j$ and $\Sigma = \partial E - \partial^* E$ be the singular parts of ∂E_j and ∂E respectively. Suppose K is a compact set in Ω and A an open set containing $\Sigma \cap K$. Then, for j sufficiently large, $\Sigma_j \cap K \subseteq A$.*

Proof: Suppose the lemma is false. Passing possibly to a subsequence we may suppose that for every j there exists a point $x_j \in \Sigma_j \cap K$ such that $x_j \notin A$. Since K is compact, we may also assume $x_j \to \bar{x} \in K$. Since $x_i \in \partial E_i$, we have, by (5.16), for every $\rho < \text{dist } (K, \partial \Omega)$

$$\int\limits_{B(x_i, \rho)} |D\varphi_{E_i}| \geq \omega_{n-1} \rho^{n-1}.$$

Now, if we choose any $r < \text{dist } (K, \partial \Omega)$ and any $\rho < r$, then for sufficiently large i

$$B(x_i, \rho) \subseteq B(\bar{x}, r).$$

Hence, as $\int\limits_{B(\bar{x},r)} |D\varphi_{E_i}|$ converges to $\int\limits_{B(\bar{x},r)} |D\varphi_E|$ for almost all r, we must have that

$$\int\limits_{B(\bar{x},r)} |D\varphi_E| \geq \omega_{n-1} r^{n-1} > 0$$

for almost all $r < \text{dist } (K, \partial \Omega)$ and so $\bar{x} \in \partial E$. Furthermore $\bar{x} \in \Sigma$ since otherwise, by Theorem 9.4, we would have $x_j \in \partial^* E_j$ for j large which contradicts our assumption on the x_j. In conclusion we must have $\bar{x} \in \Sigma \cap K \subseteq A$ which is a contradiction as A is open.   □

The next lemma concerning H^∞_k measure is now a fairly easy consequence of the last lemma.

11.5   Lemma: *Suppose $\{E_j\}$ is a sequence of minimal sets in Ω converging in $L^1_{\text{loc}}(\Omega)$ to a minimal set E. Let Σ_j and Σ be as in Lemma 11.4. Then for every compact set $K \subseteq \Omega$*

(11.6) $H_k^\infty(\Sigma \cap K) \geq \limsup\limits_{j \to \infty} H_k^\infty(\Sigma_j \cap K).$

Proof: For $\varepsilon > 0$, let $\{S_j\}$ be a covering of $\Sigma \cap K$ such that

$$H_k^\infty(\Sigma \cap K) > \omega_k 2^{-k} \sum_{j=1}^{\infty} (\text{diam } S_j)^k - \varepsilon.$$

We may suppose that the S_j are open. Let $A = \bigcup\limits_{j=1}^{\infty} S_j$. For h large enough, by Lemma 11.4 we have $\Sigma_h \cap K \subseteq A$ and hence

$$H_k^\infty(\Sigma_h \cap K) \leq H_k^\infty(A) \leq \omega_k 2^{-k} \sum_{j=1}^{\infty} (\text{diam } S_j)^k.$$

Thus

$$\limsup\limits_{h \to \infty} H_k^\infty(\Sigma_h \cap K) \leq H_k^\infty(\Sigma \cap K) + \varepsilon$$

and (11.6) follows as $\varepsilon > 0$ is arbitrary. $\square$

We now apply the previous lemma in the special case where we perform a "blow up" at some point of a minimal surface. The sets E_j will be different stages in the "blow up" and E will be the resulting minimal cone. We show that, given a minimal set with singularities of Hausdorff dimension at least k, we can find a minimal cone with singularities of the same dimension. Thus once again the regularity theory is reduced to the existence of minimal cones.

11.6 Theorem: *Suppose E is a minimal set in Ω such that $H_k(\Sigma) = H_k(\partial E - \partial^* E) > 0$. Then there exists a minimal cone C in $\mathbb{R}^n$ such that*

$$H_k(\partial C - \partial^* C) > 0.$$

Proof: From Proposition 11.3 it follows that there exists a point x_0 and a sequence $\{r_j\} \to 0$ such that

(11.7) $H_k^\infty(\Sigma \cap B(x_0, r_j)) \geq 2^{-k-1} \omega_k r_j^k.$

We may suppose $x_0 = 0$. If we set as usual

$$E_t = \{x \in \mathbb{R}^n : tx \in E\},$$

then from Theorem 9.3 there must exist a subsequence $\{s_j\}$ of $\{r_j\}$ such that E_{s_j} converges locally in $\mathbb{R}^n$ to a minimal cone C.

If we set $\Sigma_j = \partial E_{s_j} - \partial^* E_{s_j}$, then obviously

$$\Sigma_j = \{x \in \mathbb{R}^n : s_j x \in \Sigma\}$$

and hence

$$H_k^\infty(\Sigma_j \cap B_1) = s_j^{-k} H_k^\infty(\Sigma \cap B_{s_j}) \geqq 2^{-k-1} \omega_k.$$

The conclusion follows from Lemmas 11.5 and 11.2. $\square$

11.7 Theorem: *Let E be a minimal set in $\Omega \subset \mathbb{R}^8$. Then the singular set Σ consists at most of isolated points.*

Proof: Suppose on the contrary that there exists a sequence of singular points converging to some $x_0 \in \Omega$. We may suppose that $x_0 = 0$. Let $r_k = |x_k|$. From Theorem 9.3 we conclude, passing possibly to a subsequence, that the sequence E_{r_k} converges locally in $\mathbb{R}^8$ to a minimal cone C. Moreover, we may assume that $y_k = x_k/r_k$ converge to the point $y_0 = (0, 0, \ldots, 1)$.

Since y_k is a singular point for E_{r_k} we conclude from Theorem 9.4 that y_0 is a singular point for C. The set C being a cone, the whole line joining 0 with y_0 consists of singular points for C.

We now blow up near y_0 as in Proposition 9.6. Arguing as in Theorem 11.6 we obtain a minimal cylinder $Q = A \times \mathbb{R}$ such that

$$H_1(\partial Q - \partial^* Q) > 0.$$

Furthermore, by Proposition 9.9, A is a minimal cone in $\mathbb{R}^7$ and since $\partial Q - \partial^* Q = (\partial A - \partial^* A) \times \mathbb{R}$ we have

$$H_0(\partial A - \partial^* A) > 0,$$

contradicting Theorem 10.11. $\square$

When $n > 8$, using an inductive procedure as in Theorem 9.10 we can prove

11.8   Theorem (*Dimension of the singular set,* [FH3]): *Suppose that E is a minimal set in* $\Omega \subset \mathbb{R}^n$ *and let* $\Sigma = (\partial E - \partial^* E) \cap \Omega$. *Then*

$$H_s(\Sigma) = 0 \qquad \text{for all } s > n - 8.$$

Proof: We proceed as above. Let $k > 0$ be such that $H_k(\Sigma) > 0$. By Theorem 11.6 we may construct a minimal cone C in $\mathbb{R}^n$ such that $H_k(\partial C - \partial^* C) > 0$. By Proposition 11.3 we can find a point $x_0 \neq 0$ such that (11.7) holds, with Σ replaced by $\partial C - \partial^* C$. Blowing up C near x_0 we obtain as above a minimal cylinder $Q = A \times \mathbb{R}$ such that

$$H_k(\partial Q - \partial^* Q) > 0.$$

The set A is a cone and

$$H_{k-1}(\partial A - \partial^* A) > 0.$$

Repeating the argument we conclude that for every $m \leq k$ there exists a minimal cone C in $\mathbb{R}^{n-m}$ such that

$$H_{k-m}(\partial C - \partial^* C) > 0.$$

As minimal cones in $\mathbb{R}^8$ can have at most one singular point, this implies at once that $k \leq n - 8$.   □

Theorems 11.7 and 11.8 are in some sense the best possible, since the Simons cone:

$$S = \{x \in \mathbb{R}^8 : x_1^2 + \cdots + x_4^2 > x_5^2 + \cdots + x_8^2\}$$

is minimal in $\mathbb{R}^8$, as we will show later (Theorem 16.4).

The cone S provides a counterexample to interior regularity of minimal hypersurfaces. However the situation is quite different at the boundary. Indeed, Hardt and Simon [HS] have proved, in the framework of the theory of integral currents, that if the boundary manifold M is of class $C^{1,\alpha}$, then the solution of the Plateau problem is of class $C^{1,\alpha}$ in a neighborhood of M.

Part 2
Non-Parametric
Minimal Surfaces

12. Classical Solutions of the Minimal Surface Equation

Let $u(x)$ be function defined in some bounded open set Ω of $\mathbb{R}^n$. The area of the graph of u is given by

(12.1) $\quad \mathscr{A}(u, \Omega) = \int_\Omega \sqrt{1 + |Du|^2} \; dx$

where $Du = (D_1 u, \ldots, D_n u)$ is the gradient of u.

We shall always suppose in the following that the boundary of Ω, $\partial\Omega$, is at least Lipschitz-continuous. We shall consider the Dirichlet problem for the functional (12.1); i.e. the problem of minimizing the area amongst all functions taking prescribed values $\varphi(x)$ on $\partial\Omega$. Here the boundary datum φ is supposed to be Lipschitz-continuous or even smoother (in general $\varphi \in C^2$ will be sufficient). In the following chapters we discuss continuous (see theorem 13.6) and L^1 (see Theorem 14.5) boundary data.

We shall work in the space $C^{0,1}(\Omega)$ of Lipschitz-continuous functions in Ω; i.e. continuous functions with finite Lipschitz constant

$$|u|_\Omega = \sup \left\{ \frac{|u(x) - u(x)|}{|x - y|}; \; x, y \in \Omega, \; x \neq y \right\}$$

It is well-known that $C^{0,1}(\Omega)$ coincides with the space $W^{1,\infty}(\Omega)$, of functions with bounded distributional derivatives.

It is clear that the functional (12.1) is well defined for $u \in C^{0,1}(\Omega)$ (actually the same is true for $u \in W^{1,1}(\Omega)$). Moreover we have

(12.2) $\quad \int_\Omega \sqrt{1 + |Du|^2} \, dx = \sup \left\{ \int_\Omega (g_{n+1} + u \; \text{div} \; g) dx; \right.$

$$g = (g_1, \ldots, g_{n+1}) \in C^1_0(\Omega; \mathbb{R}^{n+1}), |g| \leqq 1 \}$$

and therefore, as in 1.9, we can prove that $\mathscr{A}(u)$ is lower semi-continuous with respect to weak L^1 convengence.

12.1   Definition: For $k > 0$ we set

$$L_k(\Omega) = \{ u \in C^{0,1}(\Omega) : |u|_\Omega \leqq k \}.$$

Moreover, if $\varphi \in C^{0,1}(\partial\Omega)$:

$$L_k(\Omega; \varphi) = \{u \in L_k(\Omega) : u = \varphi \text{ on } \partial\Omega\}$$

$$L(\Omega; \varphi) = \{u \in C^{0,1}(\Omega) : u = \varphi \text{ on } \partial\Omega\}.$$

12.2 Proposition: *Let φ be a Lipschitz-continuous function on $\partial\Omega$, and suppose that $L_k(\Omega; \varphi)$ is non-empty. Then the function $\mathscr{A}(u, \Omega)$ achieves its minimum in $L_k(\Omega, \varphi)$.*

Proof: Let $\{u_k\}$ be a minimizing sequence in $L_k(\Omega, \varphi)$. By the Ascoli-Arzelà theorem we can select a subsequence converging uniformly to a function $u \in L_k(\Omega, \varphi)$, and the result follows from the semicontinuity of the area.   □
  We shall denote by u^k the minimizing function in $L_k(\Omega; \varphi)$ (which is unique since $\mathscr{A}$ is strictly convex). In general, when k increases the value of the minimum decreases and $|u^k|_\Omega$ increases.

12.3 Proposition: *Let u^k be the minimum point for $\mathscr{A}$ in $L_k(\Omega, \varphi)$. If $|u^k|_\Omega < k$, then u^k minimizes $\mathscr{A}$ in $L(\Omega; \varphi)$.*

Proof: For $0 \leq t \leq 1$ and $v \in L(\Omega; \varphi)$ we set

$$v_t = u^k + t(v - u^k).$$

We have $v_t = \varphi$ on $\partial\Omega$ and for t small enough $|v_t| < k$, so that $\mathscr{A}(u^k, \Omega) \leq \mathscr{A}(v_t; \Omega)$. From the convexity of $\mathscr{A}$ we get

$$\mathscr{A}(u^k, \Omega) \leq \mathscr{A}(v_t, \Omega) \leq (1-t)\mathscr{A}(u^k, \Omega) + t\mathscr{A}(v, \Omega)$$

and therefore $\mathscr{A}(u^k, \Omega) \leq \mathscr{A}(v, \Omega)$.   □

  The above result tells us that to prove the existence of a minimum in $L(\Omega, \varphi)$ it is sufficient to get estimates for the Lipschitz constant of u^k.
  In order to simplify the notation we shall omit the index k; moreover we shall say briefly "u minimizes the area in $L_k(\Omega)$" instead of "$u \in L_k(\Omega)$ minimizes $\mathscr{A}$ among all v taking the same boundary values".
  It is clear that any u minimizing $\mathscr{A}$ in $L_k(\Omega)$ also gives a minimum for $\mathscr{A}$ in $L_{\tilde{k}}(\tilde{\Omega})$ for any $\tilde{\Omega} \subset \Omega$ and $\tilde{k} \leq k$ whenever $|u|_{\tilde{\Omega}} \leq \tilde{k}$.

Our main tool will be the weak maximum principle. In order to state it in suitable generality we introduce the notion of super- and sub-solution.

12.4 Definition: A function $w \in L_k(\Omega)$ is a supersolution [resp. subsolution] for $\mathscr{A}$ in $L_k(\Omega)$ if for every $v \in L_k(\Omega, w)$, with $v \geq w$ [resp. $v \leq w$] we have $\mathscr{A}(v, \Omega) \geq \mathscr{A}(w, \Omega)$.

In particular a function minimizing area in $L_k(\Omega)$ is both a super and a subsolution. Although not relevant for our purposes, the converse is also true, as it can be shown without difficulty.

12.5 Lemma (*Weak maximum principle*): *Let w and z be respectively a supersolution and a subsolution in $L_k(\Omega)$. If $w \geq z$ on $\partial\Omega$, then $w \geq z$ in $\bar{\Omega}$.*

Proof: Suppose on the contrary that

$$K = \{x \in \Omega : w(x) < z(x)\}$$

is non-empty, and let $v = \max\{z, w\}$.

We have $v \in L_k(\Omega; w)$ and $v \geq w$; therefore $\mathscr{A}(v, \Omega) \geq \mathscr{A}(w, \Omega)$ or, equivalently

$$\mathscr{A}(z; K) \geq \mathscr{A}(w, K)$$

In a similar way, taking $v = \min\{z, w\}$ we show $\mathscr{A}(w, K) \geq \mathscr{A}(z, K)$ and therefore

$$\mathscr{A}(w; K) = \mathscr{A}(z; K).$$

Since $z = w$ on ∂K and $z > w$ in K we must have $Dz \neq Dw$ in a set of positive measure. From the strict convexity of the area we have

$$\mathscr{A}\left(\frac{w + z}{2}; K\right) < \frac{1}{2}\mathscr{A}(w; K) + \frac{1}{2}\mathscr{A}(z; K) = \mathscr{A}(w; K).$$

But this is impossible since w is a supersolution in $L_k(K)$ and therefore

$$\mathscr{A}\left(\frac{w + z}{2}; K\right) \geq \mathscr{A}(w; K). \qquad \square$$

A simple consequence of the maximum principle is the following lemma.

12.6 Lemma: *Let w and z be respectively a supersolution and a subsolution in $L_k(\Omega)$. Then*

$$(12.3) \quad \sup_{x \in \Omega} [z(x) - w(x)] = \sup_{y \in \partial\Omega} [z(y) - w(y)]$$

Proof: It is sufficient to remark that for every $\alpha \in \mathbb{R}$, $w + \alpha$ is a supersolution, and that for $x \in \partial\Omega$:

$$z(x) \leqq w(x) + \sup_{y \in \partial\Omega} [z(y) - w(y)].$$

The result now follows from Lemma 12.5.   □

In particular, if u and v minimize area in $L_k(\Omega)$, (12.3) holds for $u - v$ and $v - u$ and hence

$$(12.4) \quad \sup_{\Omega} |u - v| = \sup_{\partial\Omega} |u - v|.$$

12.7 Lemma (*Reduction to boundary estimates*): *Let u minimize the area in $L_k(\Omega)$. Then*

$$(12.5) \quad |u|_\Omega = \sup \left\{ \frac{|u(x) - u(y)|}{|x - y|}; \; x \in \Omega, \; y \in \partial\Omega \right\}.$$

Proof: Let $x_1, x_2 \in \Omega$, $x_1 \neq x_2$, and let $\tau = x_2 - x_1$. The function

$$u_\tau(x) = u(x + \tau)$$

minimizes the area in $L_k(\Omega_\tau)$, where

$$\Omega_\tau = \{z \in \mathbb{R}^n : z + \tau \in \Omega\}.$$

The set $\Omega \cap \Omega_\tau$ is non-empty (it contains x_1) and both u and u_τ minimize the area in $L_k(\Omega \cap \Omega_\tau)$. From (12.4) we conclude that there exists $z \in \partial(\Omega \cap \Omega_\tau)$ such that

$$|u(x_1) - u(x_2)| = |u(x_1) - u_\tau(x_1)| \leq |u(z) - u_\tau(z)| = |u(z) - u(z + \tau)|.$$

On the other hand at least one of the points $z, z + \tau$ belongs to $\partial\Omega$. Denoting by L the right-hand side of (12.5) we have therefore

$$|u(x_1) - u(x_2)| \leq L|x_1 - x_2|$$

and (12.5) follows at once.   □

   In conclusion to prove the existence of a minimum we need only an estimate for $|u(x) - u(y)|$ when $y \in \partial\Omega$. This will be obtained by means of Lemma 12.5, comparing u with suitable super and subsolutions.
   For $x \in \Omega$ we denote by $d(x)$ the distance of x from $\partial\Omega$. Moreover we set for $t > 0$:

$$\Sigma_t = \{x \in \Omega : d(x) < t\}$$

$$\Gamma_t = \{x \in \Omega : d(x) = t\} = \partial\Sigma_t \cap \Omega.$$

12.8   Definition: Let φ be a Lipschitz-continuous function in $\partial\Omega$. An upper barrier v^+ (relative to φ) is a Lipschitz-continuous function defined in some Σ_{t_0}, $t_0 > 0$, satisfying

(12.6) $v^+ = \varphi$ on $\partial\Omega$; $v^+ \geq \sup_{\partial\Omega} \varphi$ on Γ_{t_0}

(12.7) v^+ is a supersolution in Σ_{t_0}.

Similarly, a lower barrier is a subsolution v^- in Σ_{t_0}, with $v^- = \varphi$ on $\partial\Omega$ and $v^- \leq \inf_{\partial\Omega} \varphi$ on Γ_{t_0}.

12.9   Theorem: *Let φ be a Lipschitz-continuous function on $\partial\Omega$, and suppose that there exist upper and lower barriers $v^\pm$ relative to φ.*
   Then the area function achieves its minimum in $L(\Omega; \varphi)$.

Proof: Let $Q \geq [v^\pm]_{\Sigma_{t_0}}$ and let $k > Q$.
   Let u give the minimum for $\mathscr{A}$ in $L_k(\Omega)$. The function u minimizes the area in $L_k(\Sigma_{t_0})$ and for every $x \in \Omega$:

$$\inf_{\partial\Omega} \varphi \leq u(x) \leq \sup_{\partial\Omega} \varphi.$$

In particular $v^-(x) \leq u(x) \leq v^+(x)$ on Γ_{t_0}, and therefore from Lemma 12.5:

$$v^-(x) \leq u(x) \leq v^+(x) \qquad\qquad \text{in } \Sigma_{t_0}.$$

Since $u = v^+ = v^-$ on $\partial\Omega$ we have

$$(12.8) \quad |u(x) - u(y)| \leq Q|x - y|$$

for every $x \in \Sigma_{t_0}$, $y \in \partial\Omega$.
   On the other hand if $x \in \Omega$ and $d(x) > t_0$ we have

$$|u(x) - u(y)| \leq \max \left\{ \sup_{\partial\Omega} \varphi - u(y), \, u(y) - \inf_{\partial\Omega} \varphi \right\} \leq Q t_0$$

and therefore (12.8) holds for every $x \in \Omega$. By Lemma 12.7 we have $|u|_\Omega \leq Q < k$ and the conclusion follows immediately from Proposition 12.3.   □

   We shall now investigate the conditions under which the construction of barriers is possible.
   We shall discuss only the case of upper barriers since if w is an upper barrier relative to $-\varphi$, $-w$ will be a lower barrier for φ.
   We remark first that the same argument leading to the Euler equation for minima gives a differential inequality for supersolutions.
   Let $v(x)$ be a C^2-supersolution on some open set Σ, and let $\eta \geq 0$ have compact support in Σ. The function

$$g(t) = \mathscr{A}(v + t\eta; \Sigma)$$

has a minimum at $t = 0$ and hence $g'(0) \geq 0$. This means that

$$\sum_{i=1}^{n} \int_{\Sigma} \frac{D_i v}{\sqrt{1 + |Dv|^2}} \, D_i \eta \, dx \geq 0$$

and therefore integrating by parts:

$$\sum_{i=1}^{n} D_i \frac{D_i v}{\sqrt{1 + |Dv|^2}} \leq 0 \text{ in } \Sigma.$$

More explicitly:

(12.9) $\mathscr{E}(v) = (1 + |Dv|^2)\, \Delta v - v_i v_j v_{ij} \leq 0$ in Σ,

where $v_i = \dfrac{\partial v}{\partial x_i}$, etc., $\Delta v = \overset{n}{\underset{i=1}{\Sigma}} \dfrac{\partial^2 v}{\partial x_i^2}$ and the sum over repeated indices is understood.

Conversely, inequality (12.9) implies that $g'(0) \geq 0$ and by the convexity of the area that v is a supersolution in Σ.

We shall now suppose that $\partial\Omega$ is of class C^2. This implies that the distance function $d(x)$ is of class C^2 in some Σ_{t_0}; moreover if $x \in \bar{\Omega}$ and $d(x) = t \leq t_0$, $-\Delta d(x)$ is the sum of the principal curvatures of Γ_t at x. Finally, if x moves along the normal to $\partial\Omega$ towards the interior of Ω, $\Delta d(x)$ decreases and hence if the mean curvature of $\partial\Omega$ is non-negative, (i.e. $\Delta d(x) \leq 0$ on $\partial\Omega$) we have $\Delta d \leq 0$ in Σ_{t_0}. For all these results see Appendix B.

We shall suppose that $\varphi \in C^2(\mathbb{R}^n)$, and we shall consider upper barriers of the form

$$v(x) = \varphi(x) + \psi(d(x))$$

where ψ is a C^2-function in $[0, R]$ satisfying

$$\psi(0) = 0,\ \psi'(t) \geq 1,\ \psi''(t) < 0$$

$$\psi(R) \geq L = 2 \sup_{\Omega} |\varphi|,$$

where $R < t_0$ will be determined later.

In this way conditions (12.6) are satisfied in Γ_R. We have easily

(12.10) $\mathscr{E}(v) = (1 + |D\varphi|^2)\Delta\varphi - \varphi_i\varphi_j\varphi_{ij} +$

$$+ \psi'\{2\varphi_i d_i \Delta\varphi + (1 + |D\varphi|^2)\Delta d - 2d_i\varphi_j\varphi_{ij} - \varphi_i\varphi_j d_{ij}\} +$$

$$+ \psi'^2\{\Delta\varphi + 2\varphi_i d_i \Delta d - \varphi_{ij}d_i d_j\} + \psi'^3 \Delta d +$$

$$+ \psi''\{1 + |D\varphi|^2 - (\varphi_i d_i)^2\},$$

where we have taken into account the fact that $|Dd| = 1$ and therefore $d_i d_{ij} = 0$.

Recalling that $\psi' \geq 1$ that $\psi'' < 0$ and that φ, d are C^2-functions in Σ_R we easily show

$$\mathscr{E}(v) \leq \psi'' + C\psi'^2 + \psi'^3 \Delta d.$$

Suppose now that the mean curvature of $\partial \Omega$ is non-negative. We have $\Delta d \leq 0$ in Σ_R and hence

$$\mathscr{E}(v) \leq \psi'' + C\psi'^2.$$

Choosing

$$\psi(d) = \frac{1}{c} \log(1 + \beta d)$$

we get $\mathscr{E}(v) \leq 0$ and we have to determine β and R in such a way that $\psi'(t) \geq 1$ and $\psi(R) \geq L$. We have

$$\psi'(t) = \frac{1}{c} \frac{\beta}{1 + \beta t} > \frac{1}{c} \frac{\beta}{1 + \beta R}$$

$$\psi(R) = \frac{1}{c} \log(1 + \beta R),$$

and all the conditions are satisfied provided we take $R = \beta^{-\frac{1}{2}}$ and β big enough.

In conclusion we have proved

12.10   Theorem: *Let Ω be a bounded open set in $\mathbb{R}^n$ with C^2-boundary of non-negative mean curvature, and let φ be a C^2 function in $\mathbb{R}^n$. Then the Dirichlet problem for the area functional and boundary datum φ is uniquely solvable in $C^{0,1}(\Omega)$.* □

A detailed discussion of the regularity of the solutions would require most of the theory of nonlinear elliptic partial differential equations in divergence form, a subject clearly beyond the scope of these notes.

A number of results relevant to our purposes are gathered in Appendix C.

Here we shall only remark that a Lipschitz continuous function u is a minimum for the area if and only if it is a weak solution of the equation

(12.11) div $T(Du) = 0$

where

$$(12.12) \ T_h(p) = \frac{p_h}{\sqrt{1 + |p|^2}}.$$

Moreover, if p is bounded ($|p| \leq M$), we have

$$\Lambda(M)|\xi|^2 \geq \frac{\partial T_h}{\partial p_i}(p)\xi_i\xi_h \geq \lambda(M)|\xi|^2$$

and therefore from Appendix C we obtain

12.11   Theorem: *Let $u \in C^{0,1}(\Omega)$ be a solution of equation (12.11) in Ω. Then u is analytic in Ω.* $\square$

Similar results hold for the boundary regularity. We have

12.12   Theorem: *Let $\partial\Omega$ and φ be of class $C^{k,\alpha}$, $2 \leq k$, and let u be a Lipschitz-continuous function in Ω, minimizing the area in $L(\Omega, \varphi)$. Then $u \in C^{k,\alpha}(\bar{\Omega})$. Moreover if $\partial\Omega$ and φ are C^∞ [analytic], then u is C^∞ [analytic] in $\bar{\Omega}$.*

We shall end this chapter by showing that the condition that the mean curvature of $\partial\Omega$ is nowhere negative is necessary for general solvability of the Dirichlet problem.

More precisely we shall prove that if the mean curvature of $\partial\Omega$ is negative at one point $x_0 \in \partial\Omega$, then there will exist smooth functions φ for which the area functional has no minimum in $L(\Omega; \varphi)$.

The following lemma is a variation of the maximum principle. To state it in the generality that we shall need in Section 16, we shall suppose that Ω is a connected open set, whose boundary is the union of four disjoint sets:

$$\partial\Omega = \Gamma_+ \cup \Gamma_- \cup \Gamma_0 \cup N$$

where Γ_+, Γ_-, Γ_0 are open (i.e. there exist disjoint open sets A_+, A_- and A_0 such that $\Gamma_+ = \partial\Omega \cap A_+$, etc.) and $H_{n-1}(N) = 0$.

Moreover, for $t > 0$ we define

$$\Omega_t = \{x \in \Omega : \text{dist}(x, \partial\Omega) > t\}.$$

12.13   Lemma: *Let Ω be as above, and let u and v be two functions of class $C^2(\Omega) \cap C^0(\Omega \cup \Gamma_0)$ such that:*

(i) $\operatorname{div} T(Dv) \leqq \operatorname{div} T(Du)$ *in* Ω

(ii) $u \leqq v$ *on* Γ_0

(iii) $\displaystyle \lim_{t \to 0^+} \int_{\partial \Omega_t - L} (1 - \langle v, T(Dv) \rangle) dH_{n-1} = 0$

for every open set $L \supset \Gamma_0 \cup \Gamma_-$

(iv) $\displaystyle \lim_{t \to 0^+} \int_{\partial \Omega_t - M} (1 + \langle v, T(Du) \rangle) dH_{n-1} = 0$

for every open set $M \supset \Gamma_0 \cup \Gamma_+$.
      Then:

(I)   *if* $\Gamma_0 \neq \varnothing$, $u \leqq v$ *in* Ω,

(II)   *if* $\Gamma_0 = \varnothing$, $u = v + const.$

Condition iii) [resp. iv] says that the graph of v [resp. u] tends to become vertical near Γ_+ [resp. Γ_-]. More precisely, the normal vector to the graph of v tends to $-v$ when x approaches Γ_+, and in particular $\partial v / \partial v \to + \infty$. Similarly, $\partial u / \partial v \to - \infty$ as $x \to \Gamma_-$.

Proof: We have for every non-negative function φ:

$$\int_{\Omega_t} \langle T(Dv) - T(Du), D\varphi \rangle dx = - \int_{\Omega_t} \varphi [\operatorname{div} T(Dv) - \operatorname{div} T(Du)] dx +$$

$$+ \int_{\partial \Omega_t} \varphi \langle v, T(Dv) - T(Du) \rangle dH_n^{-}{}_1 \geqq \int_{\partial \Omega_t} \varphi \langle v, T(Dv) - T(Du) \rangle dH_{n-1}$$

In particular we may take

$$\varphi = \varphi_k = \max \{0, \min (u - v, k) \}, \qquad k > 0.$$

We have

$$\langle T(Dv) - T(Du), D\varphi_k \rangle = \begin{cases} \langle T(Dv) - T(Du), D(u - v) \rangle & \text{if } 0 < u - v < k \\ 0 & \text{otherwise} \end{cases}$$

so that, recalling the definition of T, $\langle T(Dv) - T(Du), D\varphi_k \rangle \leq 0$ in Ω, and hence:

$$0 \geq \int_{\Omega_t} \langle T(Dv) - T(Du), D\varphi_k \rangle \, dx \geq \int_{\partial\Omega_t} \varphi_k \langle v, T(Dv) - T(Du) \rangle \, dH_{n-1}.$$

Suppose now that $\Gamma_0 \neq \emptyset$ and that $u < v$ on Γ_0, and let

$$A = \{x \in \Omega : u(x) < v(x)\}.$$

We have $\varphi_k = 0$ in A, and therefore

$$\int_{\partial\Omega_t} \varphi_k \langle v, T(Dv) - T(Du) \rangle \, dH_{n-1} \geq \int_{\partial\Omega_t - (A \cup A_-)} \varphi_k [\langle v, T(Dv) \rangle - 1] \, dH_{n-1}$$

$$+ \int_{\partial\Omega_t - (A \cup A_+)} \varphi_k [1 + \langle v \cdot T(Du) \rangle] \, dH_{n-1}$$

where A_+ and A_- are open sets such that $\Gamma_\pm = \partial\Omega \cap A_\pm$. On the other hand we have $A \cup A_- \supset \Gamma_0 \cup \Gamma_-$ and $A \cup A_+ \supset \Gamma_0 \cup \Gamma_+$, and hence passing to the limit as $t \to 0^+$ we obtain from (iii) and (iv):

$$\int_{\Omega} \langle T(Dv) - T(Du), D\varphi_k \rangle \, dx = 0.$$

Letting $k \to \infty$ we obtain

$$(12.14) \int_{\Omega} \langle T(Dv) - T(Du), D\varphi \rangle \, dx = 0$$

with $\varphi = \max(0, u - v)$.

From (12.13) and (12.14) we see at once that $D\varphi = 0$ in Ω and since $\varphi = 0$ on Γ_0 we have $\varphi = 0$ and therefore $u \leq v$ in Ω.

If we have only the weak inequality $u \leq v$ on Γ_0, we replace v by $v + \varepsilon$ and we prove the conclusion (I) letting $\varepsilon \to 0^+$.

Finally, if $\Gamma_0 = \emptyset$, we set $\lambda = v(x_0) - u(x_0)$, for some $x_0 \in \Omega$, and writing $u + \lambda$ instead of u we repeat the argument above with $A = \emptyset$. Again we conclude that

$D\varphi = 0$ and since $\varphi(x_0) = 0$ we find $\varphi = 0$ so that $v = u + \lambda$, thus proving (II).

□

12.14 Let now $x_0 \in \partial\Omega$ and let B_R be the ball of radius R centred at x_0. Taking $\delta(x) = \text{dist}(x, B_R) = |x - x_0| - R$ in $\Omega - B_R(x_0)$ and $v = A - \psi(\delta)$ for some constant A, we get from (12.10):

$$\mathscr{E}(v) = (\psi' + \psi'^3)\Delta\delta + \psi''.$$

Choosing $\psi(\delta) = -B\delta^{\frac{1}{2}}$, we obtain

$$\mathscr{E}(v) \leq \psi'^3 \Delta\delta + \psi'' \leq \frac{B\delta^{-\frac{3}{2}}}{4} \left\{ \frac{1-n}{2\,\text{diam}(\Omega)} B^2 + 1 \right\}$$

in $\Omega - \bar{B}_R(x_0)$, since $\Delta\delta = \dfrac{n-1}{d+R} \geq \dfrac{n-1}{\text{diam}(\Omega)}$.

In conclusion we have $\mathscr{E}(v) \leq 0$ in $\Omega - \bar{B}_R(x_0)$ provided we choose $B^2 > \dfrac{2\,\text{diam}(\Omega)}{n-1}$.

Let now u be a minimum for the area in $C^{0,1}(\Omega)$. If we choose $A = \sup\limits_{\partial\Omega - B_R} u + B\,\text{diam}(\Omega)^{\frac{1}{2}}$ we have $u \leq v$ in $\partial\Omega - B_R$. On the other hand $\langle T(Dv), v \rangle = 1$ on ∂B_R, and hence we can apply lemma 12.13 with $\Gamma_- = \emptyset$ in $\Omega - \bar{B}_R$ concluding that

$$\sup_{\Omega - B_R} u \leq \sup_{\partial\Omega - B_R} u + B\,\text{diam}(\Omega)^{\frac{1}{2}}.$$

In particular

$$\sup_{\partial B_R \cap \Omega} u \leq \sup_{\partial\Omega - B_R} u + B\,\text{diam}(\Omega)^{\frac{1}{2}}.$$

Suppose now that the mean curvature of $\partial\Omega$ at x_0 is negative, and let R be so small that

$$\Delta d \geq \varepsilon_0 > 0 \qquad \text{in } \Omega \cap B_R(x_0); \qquad d(x) = \text{dist}(x, \partial\Omega).$$

We may now take again $v = \alpha - \beta d^{\frac{1}{4}}$ so that

$$\mathcal{E}(v) \leqq \psi'^3 \Delta d + \psi'' \leqq \frac{\beta d^{-\frac{3}{2}}}{4}\{-\varepsilon_0 \beta^2 + 1\} < 0$$

provided $\beta^2 > \varepsilon_0^{-1}$.

If we take

$$\alpha = \sup_{\partial B_R \cap \Omega} u + \beta \, \mathrm{diam} \, (\Omega)^{\frac{1}{4}}$$

we can apply the Lemma again and conclude that

$$(12.15) \quad \sup_{\partial\Omega \cap B_R} u \leqq \sup_{\partial\Omega - B_R} u + (B + \beta) \, \mathrm{diam} \, (\Omega)^{\frac{1}{4}}$$

From this inequality the non-existence result follows at once. Indeed, if we take a smooth function φ for which (12.15) is not satisfied the Dirichlet problem with boundary datum φ cannot have a solution. $\qquad\square$

We may illustrate the above situation with a simple example.

12.15   Example: Let A_ρ^R be the annulus

$$A_\rho^R = \{x \in \mathbb{R}^2 : \rho < |x| < R\},$$

and consider the Dirichlet problem in A_ρ^R with boundary datum

$$\varphi = \begin{cases} 0 & \text{if } |x| = R \\ M & \text{if } |x| = \rho \end{cases}$$

where M is a positive constant.

From the convexity of the area and the symmetry of φ it follows easily that if v is a solution in A_ρ^R, its spherical average

$$u(r) = \frac{1}{2\pi} \int_{|y|=1} v(ry)\,ds \qquad r = |x|$$

is a solution to the same problem.

In this case the Euler equation becomes

$$u''(r) = -\frac{1}{r}u'(r)(1 + u'(r)^2)$$

and hence

$$u(r) = c \, \log \frac{R + \sqrt{R^2 - c^2}}{r + \sqrt{r^2 - c^2}}$$

where the constant c, $0 \leq c \leq \rho$, is to be determined by means of the condition $u(\rho) = M$.

On the other hand we have

$$u(\rho) = c \, \log \frac{R + \sqrt{R^2 - c^2}}{\rho + \sqrt{\rho^2 - c^2}} \leq \rho \, \log \frac{R + \sqrt{R^2 - \rho^2}}{\rho} = M_0(R, \rho)$$

and hence the problem can be solved only if $M \leq M_0$. $\square$

13. The a priori Estimate for the Gradient

In this section we shall be concerned with smooth solutions of the minimal surface equation

$$(13.1) \quad D_i\left(\frac{D_i u}{\sqrt{1 + |Du|^2}}\right) = 0$$

in a ball $\mathcal{B} \subset \mathbb{R}^n$.

Our main purpose is to prove Theorem 13.5, which gives an estimate of the gradient of u in terms of the supremum of u.

13.1 Notation: We shall denote by B_R a ball of radius R in $\mathbb{R}^{n+1}$; by $\mathcal{B}_R$ a similar ball in $\mathbb{R}^n$. If u is a solution of (13.1) in $\mathcal{B}$ we shall denote by S its graph, $S \subset \mathcal{B} \times \mathbb{R}$. We also set

$$S_R = S \cap B_R$$

$$C_R = S \cap (\mathcal{B}_R \times \mathbb{R})$$

The normal vector to S at the point $(x, u(x))$ has components

$$v_{n+1} = (1 + |Du|^2)^{-1/2}$$

$$v_i = -v_{n+1} D_i u \qquad i = 1, \ldots, n$$

From (13.1) we have

$$\delta_i v_i = 0$$

and from Lemma 10.7:

$$\mathscr{D} v_i + c^2 v_i = 0 \qquad i = 1, \ldots, n+1.$$

Setting $w = \log \dfrac{1}{v_{n+1}} = \log \sqrt{1 + |Du|^2}$ we get

(13.2) $\mathscr{D}w = |\delta w|^2 + c^2 \geqq |\delta w|^2 \geqq 0.$

13.2   Theorem: *Let u be a smooth solution of the minimal surface equation in* $\mathscr{B}_R(x_0)$, *and let* $z_0 = (x_0, u(x_0))$. *Then*

(13.3) $w(x_0) \leqq \dfrac{1}{\omega_n R^n} \displaystyle\int_{S_R(z_0)} w dH_n$

where ω_n *is the measure of the n-dimensional unit ball.*

Proof: [TN1] We can suppose $z_0 = 0$. For $0 < \varepsilon < R$ we set

$$
\varphi_\varepsilon(z) = \begin{cases}
\dfrac{1}{2(n-2)}(\varepsilon^{2-n} - R^{2-n}) + \dfrac{1}{2n}(R^{-n} - \varepsilon^{-n})|z|^2 & 0 \leq |z| < \varepsilon \\[2mm]
\dfrac{|z|^{2-n}}{n(n-2)} + \dfrac{1}{2n}|z|^2 R^{-n} - \dfrac{1}{2(n-2)}R^{-n} & \varepsilon \leq |z| \leq R \\[2mm]
0 & |z| > R
\end{cases}
$$

We have $\varphi_\varepsilon \geqq 0$ in B_R, $\varphi_\varepsilon = D\varphi_\varepsilon = 0$ on ∂B_R and hence

$$\int_S w\mathscr{D}\varphi_\varepsilon dH_n = \int_S \varphi_\varepsilon \mathscr{D}w dH_n \geqq 0.$$

From

$$\mathscr{D}|z|^\alpha = \alpha(\alpha-2)|z|^{\alpha-2}\left(1 - \frac{\langle z, v\rangle}{|z|^2}\right) + \alpha n |z|^{\alpha-2}$$

we get

$$
\mathscr{D}\varphi_\varepsilon = \begin{cases}
R^{-n} - \varepsilon^{-n} & 0 \leq |z| < \varepsilon \\
R^{-n} - |z|^{-2-n}\langle z, v\rangle^2 & \varepsilon < |z| < R \\
0 & |z| > R
\end{cases}
$$

and therefore

$$0 \leq \int_{S_\varepsilon} (R^{-n} - \varepsilon^{-n})w dH_n + \int_{S_R - S_\varepsilon} (R^{-n} - |z|^{-2-n}\langle z, v\rangle^2)w dH_n \leqq$$

$$\leq R^{-n} \int_{S_R} w \, dH_n - \varepsilon^{-n} \int_{S_\varepsilon} w \, dH_n$$

If we let $\varepsilon \to 0$ we have

$$w(0) = \lim_{\varepsilon \to 0} \frac{1}{\omega_n \varepsilon^n} \int_{S_\varepsilon} w \, dH_n$$

and the theorem is proved. $\square$

13.3   Remark: Suppose u is a solution of (13.1) in $\mathscr{B}_{2R}$ with $u(0) = 0$. Let $z_0 \in S_R$. Since $S_R(z_0) \subset S_{2R}$ we have from (13.3):

$$w(x_0) \leq c_1 R^{-n} \int_{S_{2R}} w \, dH_n$$

and therefore

$$(13.4) \quad \sup_{S_R} w(x) \leq c_1 R^{-n} \int_{S_{2R}} w \, dH_n. \qquad\qquad \square$$

13.4   Theorem: *Let u be a solution of the minimal surface equation in $\mathscr{B}_{3R}$ with $u(0) = 0$. Then*

$$(13.5) \quad R^{-n} \int_{S_R} w \, dH_n \leq c_2 \left\{ 1 + R^{-1} \sup_{\mathscr{B}_{3R}} u \right\}.$$

Proof: Let $\eta \in C_0^\infty(\mathscr{B}_{2R})$, $0 \leq \eta \leq 1$, $\eta = 1$ in $\mathscr{B}_R$ and $|D\eta| \leq 2R^{-1}$, and let

$$u_R = \begin{cases} 2R & \text{if} \quad u \geq R \\ u + R & \text{if} \quad |u| < R \\ 0 & \text{if} \quad u < -R. \end{cases}$$

We have

$$\delta_{n+1} u_R = \begin{cases} 0 & \text{if} \quad |u| > R \\ 1 - v_{n+1}^2 & \text{if} \quad |u| < R \end{cases}$$

and therefore, since

$$\int_S \delta_{n+1}(u_R\eta w)dH_n = 0$$

we get

$$\int_{S\cap\{|z_{n+1}|<R\}} (1 - v_{n+1}^2)\eta w dH_n + \int_S u_R(w\delta_{n+1}\eta + \eta\delta_{n+1}w)dH_n = 0.$$

Taking into account that $wv_{n+1} = we^{-w} \le e^{-1}$ we find

$$\int_{S\cap\{|z_{n+1}|<R\}} \eta w dH_n \le e^{-1}\int_S \eta v_{n+1}dH_n + 2R\int_{S\cap\{z_{n+1}>-R\}} w|\delta_{n+1}\eta|dH_n +$$

$$+ 2R\int_{S\cap\{z_{n+1}>-R\}} \eta|\delta w|dH_n$$

where we have used the inequality $u_R \le 2R$.

The first integral on the right-hand side can be easily estimated by remarking that

(13.6) $\int_S \eta v_{n+1}dH_n \le \int_{\mathscr{B}_{2R}} \eta dx \le \omega_n(2R)^n.$

Now let $\tau(t)$ be a smooth function with support in the interval $[-2R, \sup_{\mathscr{B}_{2R}} u + R]$, with $0 \le \tau \le 1$, $\tau \equiv 1$ in $[-R, \sup_{\mathscr{B}_R} u]$ and $|\tau'| \le 2/R$. Let $\varphi(z) = \eta(x)\tau(z_{n+1})$, and note that since η is independent of z_{n+1} we have

$$|\delta_{n+1}\eta| = |v_{n+1}v_hD_h\eta| \le v_{n+1}|D\eta| \le 2v_{n+1}R^{-1}.$$

Then

(13.7) $2R\int_{S\cap\{z_{n+1}>-R\}} w|\delta_{n+1}\eta|dH_n \le 4\int_{C_{2R}\cap\{z_{n+1}>-R\}} wv_{n+1}dH_n \le$

$$\le 4e^{-1}H_n(S\cap\mathrm{spt}\varphi)$$

and

(13.8) $\int_{S\cap\{z_{n+1}>-R\}} \eta|\delta w|dH_n \le \int_S \varphi|\delta w|dH_n \le H_n(S\cap\mathrm{spt}\,\varphi)^{1/2}(\int_S \varphi^2|\delta w|^2dH_n)^{1/2}$

In order to estimate the last integral we use again (13.2):

$$\int_S \varphi^2 |\delta w|^2 dH_n \leq \int_S \varphi^2 \mathscr{D}w dH_n \leq 2 \int_S \varphi |\delta w| \, |\delta \varphi| dH_n$$

and hence

(13.9) $(\int_S \varphi^2 |\delta w|^2 dH_n)^{1/2} \leq 2(\int_S |\delta \varphi|^2 dH_n)^{1/2} \leq 4R^{-1} H_n(S \cap \operatorname{spt} \varphi)^{1/2}$

In conclusion we have from (13.6) ... (13.9)

(13.10) $\int_{S_R} w dH_n \leq \int_{S \cap \{|z_{n+1}| < R\}} \eta w dH_n \leq c_3 \{R^n + H_n(S \cap \operatorname{spt} \varphi)\}.$

It remains to estimate $H_n(S \cap \operatorname{spt} \varphi)$. For that, let $\gamma(x)$ be a function with support in $\mathscr{B}_{3R}$, $0 \leq \gamma \leq 1$, $\gamma = 1$ in $\mathscr{B}_{2R}$ and $|D\gamma| \leq 2R^{-1}$. We have

$$\int_S \delta_{n+1} \{\gamma \max(u + 2R, 0)\} dH_n = 0,$$

and therefore

$$\int_{S \cap \{z_{n+1} > -2R\}} \gamma(1 - v_{n+1}^2) dH_n + \int_S \max(u - 2R, 0) \delta_{n+1} \gamma dH_n = 0$$

Arguing as above we find that since $\operatorname{spt} \tilde{\varphi} \subset \mathscr{B}_{2R} \times (-2R, \infty)$:

$$H_n(S \cap \operatorname{spt} \varphi) \leq \int_{C_{3R}} v_{n+1}^2 dH_n + \sup_{\mathscr{B}_{3R}} u \int_{C_{3R}} v_{n+1} |D\gamma| dH_n \leq$$

$$\leq c_4 R^n \left(1 + R^{-1} \sup_{\mathscr{B}_{3R}} u\right)$$

and the conclusion follows from (13.10).   □

Combining 13.3 and 13.4, and recalling that $w = \log \sqrt{1 + |Du|^2}$ we have proved the following

13.5 Theorem [BDGM] (*A priori estimate of the gradient*): Let u be a solution of the minimal surface equation in $\mathscr{B}_R$. Then

$$(13.11) \quad \sup_{S_{R/6}(x_0)} |Du| \leq \exp\left\{c_5\left(1 + \frac{\sup\limits_{\mathscr{B}_R(x_0)} u - u(x_0)}{R}\right)\right\}$$

In particular $|Du(x_0)|$ is bounded by the right-hand side of (13.11).

The estimate (13.11) can be put in a slightly different form. First of all we may change u in $-u$, getting the symmetric estimate

$$(13.12) \quad \sup_{S_{R/6}(x_0)} |Du| \leq \exp\left\{c_5\left(1 + \frac{u(x_0) - \inf\limits_{\mathscr{B}_R(x_0)} u}{R}\right)\right\}$$

If now u is non-negative in $\mathscr{B}_R(x_0)$, we have

$$(13.13) \quad \sup_{S_{R/6}(x_0)} |Du| \leq \exp\left\{c_5\left(1 + \frac{u(x_0)}{R}\right)\right\}.$$

Theorem 13.5 is one of the keystones of the theory of non-parametric minimal surfaces, as will be made clear in the following sections.

As a first consequence we deduce now the existence of solution to the Dirichlet problem with continuous boundary data.

13.6   Theorem: *Let Ω be a bounded open set in $\mathbb{R}^n$, with C^2-boundary $\partial\Omega$ of non-negative mean curvature, and let φ be a continuous function on $\partial\Omega$. Then the Dirichlet problem for the minimal surface equation:*

$$(13.14) \quad \sum_{i=1}^{n} D_i\left(\frac{D_i u}{\sqrt{1 + |Du|^2}}\right) = 0 \qquad \text{in } \Omega$$

$$(13.15) \quad u = \varphi \qquad\qquad\qquad \text{on } \partial\Omega$$

has a solution $u \in C^2(\Omega) \cap C^\circ(\bar{\Omega})$.

Proof: Let $\{\varphi_j\}$ be a sequence of C^2-functions in $\mathbb{R}^n$, converging uniformly to φ in $\partial\Omega$.

Let u_j be the function minimizing the area in $L(\Omega, \varphi_j)$ (Theorem 12.10). We know that the u_j are regular in Ω and satisfy (13.14).

We have from (12.4)

$$\sup_{\Omega} |u_j - u_h| \leq \sup_{\partial\Omega} |\varphi_j - \varphi_h|$$

and therefore u_j converge uniformly in Ω to some function u. Now let K be a compact set, $K \subset \Omega$.

From Theorem 13.5 we get easily

$$\sup_K |Du_j| \leqq L$$

where L depends on K but not on j. Moreover the theory of uniformly elliptic equations (see Theorem C10) gives a bound for the derivatives of any order:

$$\sup_K |D^s u_j| \leqq L(K, s)$$

again with L independent of j.

In particular we may conclude that $u \in C^2(\Omega) \cap C^0(\bar{\Omega})$ and that $u_j \to u$ in $C^2_{loc}(\Omega)$ so that u is a solution of the Dirichlet problem (13.14), (13.15). $\square$

If the boundary datum φ is more than merely continuous, the solution u is correspondingly more regular. We shall only state here the results, referring for the proofs to the original paper.

13.7   Theorem [GE2]: *Let $\partial \Omega$ be of class C^2 and let u be the solution of the Dirichlet problem* (13.14), (13.15). *Then:*

(i) *if $\varphi \in C^{0,\alpha}(\partial \Omega)$ and the mean curvature of $\partial \Omega$ is strictly positive, $u \in C^{0,\alpha/2}(\bar{\Omega})$.*
(ii) *if $\varphi \in C^{1,\alpha}(\partial \Omega)$, and the mean curvature of $\partial \Omega$ is non-negative, then $u \in C^{0,1}(\bar{\Omega})$.*

We remark that (i) is sharp, and we may have $\varphi \in C^{0,1}$ with $u \in C^{0,1/2}$. To see that, let us come back to Example 12.15. The function:

$$u(x) = \rho \log \frac{R + \sqrt{R^2 - \rho^2}}{r + \sqrt{r^2 - \rho^2}}, \qquad r = |x|$$

is a solution of the minimal surface equation in the amulus $A_\rho^R = \{x \in \mathbb{R}^2, \rho < |x| < R\}$.

If B is a disc contained in A_ρ^R and tangent to the internal circle, we have $u \in C^{0,1/2}(B)$ whereas the restriction of u to ∂B is Lipschitz-continuous.

Some additional comments are needed at this point, concerning the

equivalence of the Dirichlet problem (13.14), (13.15) with the least area problem
with boundary datum φ.

As usual, we shall suppose that $\partial\Omega$ has Lipschitz-continuous boundary.

It is clear that if $u \in C^{0,1}(\Omega) \cap C^0(\bar{\Omega})$ minimizes the area among all functions
taking the values $\varphi(x)$ at $\partial\Omega$, it is smooth in Ω and satisfies the Euler equation
(13.14).

We shall show that the converse is also true, namely if u is a solution to the
Dirichlet problem, the area of the graph of u is finite, and actually u minimizes
the area integral.

Suppose first that $\mathscr{A}(u, \Omega) < +\infty$, and let v be any function with $v = \varphi$ on $\partial\Omega$.
We have, since $u - v = 0$ on $\partial\Omega$:

$$\int_\Omega \frac{D_i u D_i (u - v)}{\sqrt{1 + |Du|^2}} \, dx = 0$$

and hence

$$\int_\Omega \sqrt{1 + |Du|^2} \, dx = \int_\Omega \frac{1 + |Du|^2}{\sqrt{1 + |Du|^2}} \, dx =$$

$$= \int_\Omega \frac{1 + D_i u D_i v}{\sqrt{1 + |Du|^2}} \, dx \leqq \int_\Omega \sqrt{1 + |Dv|^2} \, dx$$

so that u minimizes the area.

To prove that $\mathscr{A}(u, \Omega) < +\infty$ we observe that for any $\delta > 0$, $\mathscr{A}(u, \Omega_\delta) < +\infty$
and therefore u minimizes the area in $\Omega_\delta = \{x \in \Omega: \text{dist}\,(x, \partial\Omega) > \delta\}$. Now let

$$\eta(t) = \begin{cases} 2 - t/\delta & \delta \leqq t < 2\delta \\ 0 & t \geqq 2\delta \end{cases}$$

and let

$$v(x) = u(x)\eta(d(x))$$

we have $v = u$ on $\partial\Omega_\delta$ and therefore

$$\mathscr{A}(u, \Omega_\delta) \leqq \mathscr{A}(v, \Omega_\delta).$$

Now

$$\mathscr{A}(v,\Omega_\delta) \leqq |\Omega_\delta| + \int_{\Omega_\delta} \eta\,|Du|\,dx + \int_{\Omega_\delta} |u|\,|\eta'|\,dx \leqq$$

$$\leqq |\Omega| + \int_{\Omega_\delta - \Omega_{2\delta}} |Du|\,dx + \frac{1}{\delta}\sup_\Omega |u|\,|\Omega_\delta - \Omega_{2\delta}|$$

and in conclusion

$$\int_{\Omega_{2\delta}} \sqrt{1+|Du|^2} \leqq |\Omega| + c_1 \sup_\Omega |u|.$$

Since the right-hand side is independent of δ we may conclude that u has finite area in Ω.

We have then proved

13.8 Theorem: *A function $u \in C^2(\Omega)\cap C^0(\bar\Omega)$ is a solution of the Dirichlet problem (13.14), (13.15) if and only if it minimizes the area among all functions taking boundary values φ on $\partial\Omega$. In particular the solution of the Dirichlet problem is unique.*

We remark that Theorem 13.8 is a peculiarity of the minimal surface equation, and it is not true for instance for harmonic functions which may have infinite Dirichlet integral.

14. Direct Methods

An alternative approach to the existence of non-parametric minimal surfaces with prescribed boundary data consists in using direct methods in the calculus of variations to minimize the area integrand

$$(14.1) \quad \mathscr{A}(u; \Omega) = \int_{\Omega} \sqrt{1 + |Du|^2} \, dx$$

among all the functions taking prescribed values $\varphi(x)$ on $\partial\Omega$. As for the parametric case, the natural space here is $BV(\Omega)$, the space of functions whose derivatives are Radon measures in Ω.

14.1 Definition: Let Ω be an open set in $\mathbb{R}^n$. for $f \in BV(\Omega)$ we define

$$\int_{\Omega} \sqrt{1 + |Df|^2} = \sup\left\{ \int_{\Omega} (g_{n+1} + fD_i g_i) dx; \right.$$

$$\left. g = (g_1, \ldots, g_{n+1}) \in C_0^1(\Omega); |g| \leq 1 \right\}.$$

It is easily seen that

$$(14.2) \quad \int_{\Omega} |Df| \leq \int_{\Omega} \sqrt{1 + |Df|} \leq \int_{\Omega} |Df| + |\Omega|;$$

moreover, if $f \in W^{1,1}(\Omega)$ we have as in Example 1.2:

$$\int_{\Omega} \sqrt{1 + |Df|^2} = \int_{\Omega} \sqrt{1 + |\text{grad } f|^2} \, dx.$$

Many of the results of chapter 1 hold with $|Df|$ replaced by $\sqrt{1 + |Df|^2}$. We shall not repeat here the proofs, which are simple variations of those given there. We only state the semicontinuity theorem.

14.2 Theorem: *Let u_j converge to u in $L_{loc}^1(\Omega)$. Then*

$$(14.3) \quad \int_{\Omega} \sqrt{1 + |Du|^2} \leq \liminf_{j \to \infty} \int_{\Omega} \sqrt{1 + |Du_j|^2}$$

The direct method consists in minimizing the integral (14.1) among all $u \in BV(\Omega)$ whose trace on $\partial\Omega$ is a prescribed function $\varphi \in L^1(\partial\Omega)$ (as usual we suppose Ω bounded and $\partial\Omega$ Lipschitz-continuous).

It is easily seen from (14.2) and Theorem 1.19 that minimizing sequences are relatively compact in $L^1(\Omega)$, and hence from any minimizing sequence we can extract a subsequence u_j converging in $L^1(\Omega)$ to some function u. We can now apply (14.3), but since we do not know that u has trace φ on $\partial\Omega$ we cannot conclude that u gives the required minimum. On the other hand, the non-existence results of section 12 suggest that even in this more general setting the Dirichlet problem might not be solvable for general domains Ω.

14.3   Proposition: *Let Ω be a bounded open set with C^1 boundary $\partial\Omega$, and let $\varphi \in L^1(\partial\Omega)$. We have*

(14.4) $\inf\{\mathscr{A}(u; \Omega) : u \in BV(\Omega), \, u = \varphi \text{ on } \partial\Omega\} =$

$$= \inf\left\{\mathscr{A}(u; \Omega) + \int_{\partial\Omega} |u - \varphi| \, dH_{n-1}; \, u \in BV(\Omega)\right\}.$$

Proof: It is sufficient to show that the left-hand side of (14.4) is not greater than its right-hand side.

Let $u \in BV(\Omega)$ and let $\varepsilon > 0$. From Theorem 2.16 and Remark 2.17 it follows that there exists a function $w \in W^{1,1}(\Omega)$, with $w = \varphi - u$ on $\partial\Omega$ and

$$\int_{\Omega} |Dw| \, dx \le (1 + \varepsilon) \int_{\partial\Omega} |u - \varphi| \, dH_{n-1}.$$

The function $v = u + w$ is in $BV(\Omega)$, and $v = \varphi$ on $\partial\Omega$. Moreover

$$\int_{\Omega} \sqrt{1 + |Dv|^2} \le \int_{\Omega} \sqrt{1 + |Du|^2} + \int_{\Omega} |Dw| \, dx \le$$

$$\le \int_{\Omega} \sqrt{1 + |Du|^2} + (1 + \varepsilon) \int_{\Omega} |u - \varphi| \, dH_{n-1}$$

from which (14.4) follows at once.   □

The above result suggest the following weaker form of the Dirichlet problem.

14.4   Dirichlet Problem for Non-parametric Minimal Surfaces

Given a function φ in $L^1(\partial\Omega)$, find a function u in $BV(\Omega)$ minimizing the functional

$$(14.5) \quad \mathscr{I}(v;\Omega) = \int_{\Omega}\sqrt{1+|Dv|^2} + \int_{\partial\Omega}|v-\varphi|dH_{n-1}$$

among all function $v \in BV(\Omega)$.   □

As we shall see shortly, the problem 14.4 has always a solution. We remark that the last integral in (14.5) represents the area of that part of the cylinder $\partial\Omega \times \mathbb{R}$ lying between the graphs of φ and u, and hence it can be seen as a penalization for not taking the boundary values φ on $\partial\Omega$. Let Ω be a bounded open set with Lipschitz-continuous boundary. If $\mathscr{B}$ is a ball containing $\bar{\Omega}$ we can use Theorem 2.16 to extend φ to a $W^{1,1}$ function in $\mathscr{B} - \bar{\Omega}$, that we will denote again by φ. If we set for $v \in BV(\Omega)$

$$v_\varphi(x) = \begin{cases} v(x) & x \in \Omega \\ \varphi(x) & x \in \mathscr{B} - \Omega \end{cases}$$

the function v_φ belongs to $BV(\mathscr{B})$ and by (2.15)

$$\int_{\mathscr{B}}\sqrt{1+|Dv_\varphi|^2} = \int_{\Omega}\sqrt{1+|Dv|^2} + \int_{\mathscr{B}-\Omega}\sqrt{1+|D\varphi|^2}\,dx + \int_{\partial\Omega}|v-\varphi|dH_{n-1}$$

$$= \mathscr{I}(v,\Omega) + \int_{\mathscr{B}-\Omega}\sqrt{1+|D\varphi|^2}\,dx$$

We have therefore an equivalent formulation of the Dirichlet problem: Given a function $\varphi \in W^{1,1}(\mathscr{B} - \bar{\Omega})$ find a function $u \in BV(\mathscr{B})$, coinciding with φ in $\mathscr{B} - \bar{\Omega}$ and minimizing the area $\mathscr{A}(v;\mathscr{B})$ among all function $v \in BV(\mathscr{B})$ with $v = \varphi$ in $\mathscr{B} - \bar{\Omega}$.

We can now apply Theorems 1.19 and 14.2 and conclude at once the existence of a minimum. We have therefore:

14.5   Theorem: *Let Ω be a bounded open set with Lipschitz-continuous boundary $\partial\Omega$, and let φ be a function in $L^1(\partial\Omega)$. Then the functional*

$$\mathscr{I}(u;\Omega) = \int_{\Omega}\sqrt{1+|Du|^2} + \int_{\partial\Omega}|u-\varphi|dH_{n-1}$$

attains its minimum in $BV(\Omega)$.

We shall prove now that any function u minimizing $\mathscr{I}$ is regular (analytic) in Ω. To do this we show that the subgraph of u is a set of least perimeter, so that we can use the regularity results of the first part. For that first we need some results connecting non-parametric and parametric minimal surfaces.

14.6   Theorem: *Let* $u \in BV(\Omega)$ *and let*

$$U = \{(x, t) \in \Omega \times \mathbb{R} : t < u(x)\}$$

be the subgraph of u. Then

$$(14.6) \quad \int_{\Omega} \sqrt{1 + |Du|^2} = \int_{\Omega \times \mathbb{R}} |D\varphi_U|$$

Proof: Suppose first that u is bounded. By adding a constant we may suppose that $u \geq 1$.

Let $g_1(x), \ldots, g_{n+1}(x)$ be functions with compact support in Ω satisfying $|g| \leq 1$ in Ω, and let $\eta(t)$ be a function with support in $[0, \sup_{\Omega} u + 1]$ such that $\eta \equiv 1$ in $[1, \sup_{\Omega} u]$, and $|\eta| \leq 1$.

Let $\gamma_i(x, x_{n+1}) = g_i(x)\eta(x_{n+1})$, $i = 1, \ldots, n+1$; we have $|\gamma| \leq 1$ in $\Omega \times \mathbb{R}$ and therefore

$$\int_{\Omega \times \mathbb{R}} |D\varphi_U| \geq \int_U \sum_{i=1}^{n+1} D_i\gamma_i \, dx \, dx_{n+1} =$$

$$= \int_{\Omega} dx \int_0^{u(x)} [g_{n+1}\eta'(x_{n+1}) + \eta(x_{n+1}) \sum_{i=1}^{n} D_i g_i(x)] dx_{n+1}$$

Now $\displaystyle\int_0^{u(x)} \eta'(x_{n+1}) dx_{n+1} = 1$

and $\displaystyle\int_0^{u(x)} \eta(x_{n+1}) dx_{n+1} = u(x) - \int_0^1 (1 - \eta(x_{n+1})) dx_{n+1} = u(x) - c.$

We have therefore

$$\int_{\Omega \times \mathbb{R}} |D\varphi_U| \geq \int_{\Omega} (g_{n+1} + u \sum_{i=1}^{n} D_i g_i) dx$$

and hence

$$\int_{\Omega \times \mathbb{R}} |D\varphi_U| \ge \int_\Omega \sqrt{1 + |Du|^2}.$$

To prove the opposite inequality, we remark first that (14.6) holds for C^1-functions, since in this case both members are equal to the area of the graph of u elementarily defined.

Now let $u_j \in C^\infty(\Omega)$, $u_j \to u$ in $L^1(\Omega)$ and $\int_\Omega \sqrt{1 + |Du_j|^2} \to \int_\Omega \sqrt{1 + |Du|^2}$. We have $U_j \to U$ in $L^1_{loc}(\Omega \times \mathbb{R})$ and therefore

$$\int_{\Omega \times \mathbb{R}} |D\varphi_U| \le \liminf_{j \to \infty} \int_{\Omega \times \mathbb{R}} |D\varphi_{U_j}| = \lim_{j \to \infty} \int_\Omega \sqrt{1 + |Du_j|^2}$$

$$= \int_\Omega \sqrt{1 + |Du|^2}.$$

Finally, writing (13.2) for

$$u_T(x) = \begin{cases} u(x) & \text{if} \quad |u| < T \\ T & \text{if} \quad u \ge T \\ -T & \text{if} \quad u \le -T \end{cases}$$

and letting $T \to +\infty$ we get the full result.   □

We now show that given a set F we can decrease its perimeter by replacing it with a suitable subgraph.

14.7 **Lemma:** *Let $F \subset Q = \Omega \times \mathbb{R}$ be a measurable set, and suppose that for some $T > 0$ we have*

$$\Omega \times (-\infty, -T) \subset F \subset \Omega \times (-\infty, T).$$

For $x \in \Omega$ let

$$w(x) = \lim_{k \to \infty} \left(\int_{-k}^{k} \varphi_F(x, t)dt - k \right).$$

Then

$$\int_{\Omega} \sqrt{1 + |Dw|^2} \leqq \int_{\Omega \times \mathbb{R}} |D\varphi_F|.$$

Proof: We have

$$\partial F \cap Q \subset \Omega \times (-T, T).$$

Setting

$$w_k = \int_{-k}^{k} \varphi_F(x, t)\,dt - k$$

and remarking that $w_h = w_k$ for $k, h \geqq T$ we conclude that $w(x)$ is a bounded measurable function, and $-T \leqq w(x) \leqq T$.

Let now $g(x) = (g_1(x), \ldots, g_{n+1}(x))$ be in $C_0^1(\Omega; \mathbb{R}^{n+1})$ and satisfy $|g| \leqq 1$ and let $\eta(t)$ be a smooth function such that $0 \leqq \eta \leqq 1$ and

$$\eta(t) = 0 \quad \text{if} \quad |t| \geqq T + 1$$

$$\eta(t) = 1 \quad \text{if} \quad |t| \leqq T.$$

We have

$$\int_{-\infty}^{\infty} \eta'(t)\varphi_F(x, t)\,dt = 1$$

$$\int_{-\infty}^{\infty} \eta(t)\varphi_F(x, t)\,dt = w(x) + T + \int_{-T-1}^{-T} \eta\,dt = w(x) + \alpha,$$

and therefore

$$\int_{Q} |D\varphi_F| \geqq \int_{Q} \varphi_F(x, x_{n+1}) \sum_{i=1}^{n+1} \frac{\partial}{\partial x_i}[\eta(x_{n+1})g_i(x)]\,dx\,dx_{n+1}$$

$$= \int_{\Omega} \{(w + \alpha) \sum_{i=1}^{n} \frac{\partial g_i}{\partial x_i} + g_{n+1}\}\,dx = \int_{\Omega} (g_{n+1} + w\,\mathrm{div}\,g)\,dx.$$

If we take the supremum over g we get the conclusion of the lemma. $\qquad\square$

We now want to remove the restriction that $\partial F \cap Q$ be bounded. We have

14.8   Theorem: *Let F be a measurable set in Q, and suppose that*

(i) For almost every $x \in \Omega$ we have

$$\lim_{t \to +\infty} \varphi_F(x, t) = 0$$

$$\lim_{t \to -\infty} \varphi_F(x, t) = 1.$$

(ii) The symmetric difference $F_0 = (F - Q^-) \cup (Q^- - F)$ has finite measure

$$(Q^- = \{(x, t) \in Q, \, t < 0\}).$$

Then the function

$$w(x) = \lim_{k \to \infty} \left(\int_{-k}^{k} \varphi_F(x, t) dt - k \right)$$

belongs to $L^1(\Omega)$, and

$$\int_{\Omega} \sqrt{1 + |Dw|^2} \leqq \int_Q |D\varphi_F|.$$

Remark: We observe that our hypotheses are redundant; in particular (ii) follows from (i) and from the finiteness of the perimeter of F in Q, using the isoperimetric inequality (1.19).

Proof: From (i) it follows that the sequence w_k converges to w almost everywhere in Ω. Moreover

$$\int_{\Omega} |w| dx = |F_0|,$$

and since $|w_k| \leqq |w|$ we have $w_k \to w$ in $L^1(\Omega)$.

Now let

$$F_k = F \cup [\Omega \times (-\infty, \, -k)] - [\Omega \times (k, \, +\infty)].$$

From the preceding Lemma we get

$$\int_\Omega \sqrt{1 + |Dw_k|^2} \le \int_Q |D\varphi_{F_k}| \le \int_Q |D\varphi_F| + \int_{\Omega \times \{k\}} \varphi_F dH_n + \int_{\Omega \times \{-k\}} (1 - \varphi_F) dH_n$$

and the conclusion follows from (i) and the lower semi-continuity of the area. $\square$

Let us now turn to non-parametric minimal surfaces. We have

14.9   Theorem: [MM2] *Let* $u \in BV_{loc}(\Omega)$ *be a local minimum of the area functional. Then the set*

$$U = \{(x, t) \in \Omega \times \mathbb{R}: t < u(x)\}$$

minimizes locally the perimeter in $Q = \Omega \times \mathbb{R}$.

Proof. Let $A \subset\subset \Omega$ and let F be any Caccioppoli set in Q, coinciding with U outside some compact set $K \subset A \times \mathbb{R}$. It is clear that U, and therefore F, satisfies condition (i) and (ii) of Theorem 14.8. The function w coincides with u outside A and hence

$$\int_{A \times \mathbb{R}} |D\varphi_U| = \int_A \sqrt{1 + |Du|^2} \le \int_A \sqrt{1 + |Dw|^2} \le \int_{A \times \mathbb{R}} |D\varphi_F|. \qquad\qquad \square$$

As a consequence of the above result we have

14.10   Theorem: *Let* $u \in BV_{loc}(\Omega)$ *minimize the area functional. Then* u *is locally bounded in* Ω.

Proof: Suppose that there exists a compact set $K \subset \Omega$ such that u is not bounded on K (for instance let $\sup_K u = +\infty$).

Let $R = \frac{1}{2}\text{dist}(K, \partial\Omega)$. For every $H \in \mathbb{N}$ there exists a point $x \in K$ such that $u(x) > 2HR$. It follows that the points z_i of $\mathbb{R}^{n+1}$, whose coordinates are $(x, 2iR)$ $(i = 0, 1, \ldots, H)$ belong to U. From Proposition 5.14 we get

$$|U \cap B(z_i, R)| \ge cR^{n+1}$$

with c depending only on n. But then

$$\int_{K_R} |u|\,dx \geq \sum_{i=1}^{H} |U \cap B(z_i, R)| \geq cHR^{n+1}$$

where $K_R = \{x \in \Omega : \text{dist}(x, K) < R\}$. Since H is arbitrary, this would imply that $\int_{K_R} |u|\,dx = +\infty$, contrary to the hypothesis. $\square$

It is easily seen that the above argument gives an estimate of $\sup_{K} |u|$ in terms of $R = \frac{1}{2}\text{dist}(K, \partial\Omega)$ and $\|u\|_{L^1(K_R)}$ only. A second consequence of Theorem 14.9 is that the boundary of U, ∂U, is a regular hypersurface outside a closed set Σ, with $H_{n-6}(\Sigma) = 0$ (see Theorem 11.8). We shall prove now that in the set

$$L = \Omega - \text{proj}\,\Sigma$$

the function u is regular. To see that, it is sufficient to show that $v_{n+1} > 0$ on $\partial U - \Sigma$. Suppose on the contrary that at a point $x_0 \in \partial U - \Sigma$ we have $v_{n+1}(x_0) = 0$. Then in a neighborhood of x_0 it is possible to represent ∂U as the graph of a smooth function:

$$x_1 = F(x_2, \ldots, x_{n+1})$$

with $\quad \dfrac{\partial F}{\partial x_{n+1}}(x_0) = 0, \quad \dfrac{\partial F}{\partial x_{n+1}} \geq 0.$

The function F is of course a solution of the minimal surface equation

$$\sum_{i=2}^{n+1} \frac{\partial}{\partial x_i}\left(\frac{\partial F/\partial x_i}{\sqrt{1 + |DF|^2}} \right) = 0$$

and therefore, as in section 12, $v = \partial F/\partial x_{n+1}$ is a solution of the uniformly elliptic equation

$$\sum_{i,j=2}^{n+1} \frac{\partial}{\partial x_i}\left\{ \frac{(1 + |DF|^2)\delta_{ij} - D_iF D_jF}{(1 + |DF|^2)^{3/2}} \frac{\partial v}{\partial x_j} \right\} = 0.$$

From the strong maximum principle (Theorem C7) we conclude that $\partial F/\partial x_{n+1} \equiv 0$ and therefore that v_{n+1} vanishes identically in a neighborhood V of x_0. Let $\Gamma = \text{proj}\,V$. We have $H_{n-1}(\Gamma) > 0$; moreover, if $z \in \Gamma$, the vertical straight line through z contains a point $x \in \partial U - \Sigma$ with $v_{n+1}(x) = 0$. It follows from the above that if this line does not meet Σ, it lies entirely on ∂U. This is of course impossible, since u is locally bounded, and therefore we must have

$$\Gamma \subset \mathrm{proj}\, \Sigma.$$

But then $H_{n-1}(\Sigma) > 0$, a contradiction. $\qquad\square$

14.11   Proposition: *Let* $u \in BV_{loc}(\Omega)$ *minimize the area in* Ω. *Then* $u \in W_{loc}^{1,1}(\Omega)$.

Proof: Let $S = \mathrm{proj}\,\Sigma$; we have seen that u is regular in $\Omega - S$ and $H_{n-6}(S) = 0$, hence in particular $|S| = 0$. If $A \subset\subset \Omega$ is an open set, we have

$$\int_A \sqrt{1 + |Du|^2} = \int_{A-S} \sqrt{1 + |Du|^2}\, dx + \int_{S \cap A} |Du|$$

since S has zero measure. On the other hand Theorem 4.4 tells us that $P(U, A \times \mathbb{R}) = H_n(\partial^* U \cap A \times \mathbb{R})$ and therefore

$$\int_A \sqrt{1 + |Du|^2} = \int_{A-S} \sqrt{1 + |Du|^2}\, dx$$

so that $\int_{S \cap A} |Du| = 0$ and $u \in W_{loc}^{1,1}(\Omega)$. $\qquad\square$

Since the area functional is strictly convex in $W^{1,1}(\Omega)$, we have easily

14.12   Proposition: *Let* Ω *be connected and let* $\varphi \in L^1(\partial\Omega)$ *and let* u, v *be two minima of the functional:*

$$\int_\Omega \sqrt{1 + |Du|^2} + \int_{\partial\Omega} |u - \varphi|\, dH_{n-1}.$$

Then $v = u + \text{const.}$

Proof: From the strict convexity we infer $Du = Dv$ and therefore $v = u + \text{const.}$ $\square$

We can prove now the regularity theorem.

14.13   **Theorem:** *Let* $u \in BV_{loc}(\Omega)$ *minimize locally the functional*

$$\int \sqrt{1 + |Du|^2}$$

Then u is Lipschitz-continuous (and hence analytic) in Ω.

Proof: Let $\mathscr{B} = \mathscr{B}(x_0, R)$ be a ball in Ω. We have

$$\int_{\mathscr{B}} \sqrt{1 + |Du|^2}\, dx \leq \int_{\mathscr{B}} \sqrt{1 + |Dw|^2} + \int_{\partial\mathscr{B}} |w - u|\, dH_{n-1}$$

for every $w \in BV(\mathscr{B})$. Since the singular set S satisfies $H_{n-6}(S) = 0$, we can find a sequence S_h of open sets such that

$$S_h \supset\supset S_{h+1} \qquad h = 1, 2, \ldots$$

$$\bigcap_{h \in N} S_h = S$$

and

$$H_{n-1}(S_j \cap \partial\mathscr{B}) \to 0$$

Let now φ_j be a smooth function on $\partial\mathscr{B}$ satisfying

$$\varphi_j = u \quad \text{in} \quad \partial\mathscr{B} - S_j$$

$$\sup_{\partial\mathscr{B}} |\varphi_j| \leq 2 \sup_{\partial\mathscr{B}} |u|,$$

and let u_j be the (unique) solution of the Dirichlet problem with datum φ_j on $\partial\mathscr{B}$ (see Theorem 12.10). The functions u_j are smooth in $\mathscr{B}$, and moreover

$$\sup_{\mathscr{B}} |u_j| \leq 2 \sup_{\partial\mathscr{B}} |u|.$$

We have

$$(14.7) \quad \int_{\mathscr{B}} \sqrt{1 + |Du_j|^2}\, dx \leq \int_{\mathscr{B}} \sqrt{1 + |Dw|^2} + \int_{\mathscr{B}} |w - \varphi_j|\, dH_{n-1}$$

for every $w \in BV(\mathscr{B})$.

From the a-priori estimate of the gradient (Theorem 13.5) we conclude that the gradients Du_j are equibounded in every compact set $K \subset \mathscr{B}$. By the Ascoli-Arzelà theorem a subsequence, that we shall again denote by u_j, will converge uniformly on compact subsets of $\mathscr{B}$ to a (locally) Lipschitz-continuous function v. Taking $w = 0$ in (14.7) we get

$$\int_{\mathscr{B}} \sqrt{1 + |Du_j|^2} \leqq |B| + \int_{\partial\mathscr{B}} |\varphi_j| dH_{n-1} \leqq \text{const.}$$

and therefore $v \in W^{1,1}(\mathscr{B})$.

We want to prove that v has trace u on $\partial\mathscr{B}$. For that, let $y \in \partial\mathscr{B}$ be a regular point for u. For j sufficiently large, $y \in \partial\mathscr{B} - S_j$ and therefore for all $k > j$ $\varphi_k = u$ in a neighborhood of y in $\partial\mathscr{B}$. We can therefore construct two functions, φ^+ and φ^-, of class C^2 on $\partial\mathscr{B}$, such that

(i) $\varphi^{\pm} = u$ in a neighborhood of y in $\partial\mathscr{B}$

(ii) $\varphi^- \leqq \varphi_k \leqq \varphi^+$ in $\partial\mathscr{B}$ for every $k > j$.

Let $u^{\pm}$ be the solutions of the Dirichlet problem with data $\varphi^{\pm}$ respectively. We have

$$u^- \leqq u_k \leqq u^+ \qquad \forall k > j$$

and therefore

$$u^- \leqq v \leqq u^+.$$

We can therefore conclude that $v = u$ at every regular point $y \in \partial\mathscr{B}$ and hence, since $H_{n-1}(S) = 0$, that v has trace u on $\partial\mathscr{B}$.

Passing now to the limit in (14.7) and remarking that $\varphi_j \to u$ in $L^1(\partial\mathscr{B})$ we have

$$(14.8) \quad \int_{\mathscr{B}} \sqrt{1 + |Dv|^2}\, dx \leqq \int_{\mathscr{B}} \sqrt{1 + |Dw|^2} + \int_{\partial\mathscr{B}} |w - u| dH_{n-1}$$

and since $v = u$ on $\partial\mathscr{B}$, the function v, as well as u, minimizes the functional on the right-hand side of (14.8). By Proposition 14.12 $v = u + \text{const}$ and since $v = u$ on $\partial\mathscr{B}$ we get $v = u$. This proves that u is Lipschitz-continuous, and hence analytic, in Ω.   □

15.   Boundary Regularity

We shall now discuss the boundary regularity of non-parametric minimal surfaces. For that we first need some lemmas.

15.1   Lemma: *Let E and F be Caccioppoli sets. Then for any open set $A \subset \mathbb{R}^n$:*

$$(15.1) \quad \int_A |D\varphi_{E \cup F}| + \int_A |D\varphi_{E \cap F}| \leq \int_A |D\varphi_E| + \int_A |D\varphi_F|.$$

Proof: Let f and g be smooth functions, with $0 \leq f \leq 1$, $0 \leq g \leq 1$, and let

$$\varphi = f + g - fg; \quad \psi = fg.$$

We have easily

$$|D\varphi| \leq (1 - f)|Dg| + (1 - g)|Df|$$

$$|D\psi| \leq f|Dg| + g|Df|$$

and therefore

$$\int_A |D\varphi| + \int_A |D\psi| \leq \int_A |Df| + \int_A |Dg|.$$

Now let f_j and g_j two sequences of smooth functions converging to φ_E, φ_F respectively, and such that $\int_A |Df_j| \to \int_A |D\varphi_E|$; $\int_A |Dg_j| \to \int_A |D\varphi_F|$ (Theorem 1.17).

Since $\varphi_j \to \varphi_{E \cup F}$ and $\psi_j \to \varphi_{E \cap F}$ we get at once (15.1) from the semi-continuity theorem 1.9. $\qquad\square$

15.2   Remark: A consequence of the above Lemma is that if F and E have least perimeter in A, and if $E \Delta F = (E - F) \cup (F - E) \subset\subset A$, then $E \cup F$ and $E \cap F$ have least perimeter in A. Indeed, $E \cup F = F \cup (E - F)$ and $E \cap F = E - (E - F)$, and therefore

$$\int |D\varphi_{E \cup F}| \geqq \int |D\varphi_F|$$

$$\int |D\varphi_{E \cap F}| \geqq \int |D\varphi_E|$$

so that by (15.1) the equality sign holds in both the above expressions.

15.3 Lemma: *Let $E = E_1 \cup E_2$ and let $H_{n-1}(\bar{E}_1 \cap \bar{E}_2) = 0$. Then for any open set A we have*

$$(15.2) \quad \int_A |D\varphi_E| = \int_A |D\varphi_{E_1}| + \int_A |D\varphi_{E_2}|$$

Moreover if E has least area in A, the same is true for E_1 and E_2.

Proof: We have

$$H_{n-1}(\partial^* E_1 \cap \partial^* E_2) = H_{n-1}(\partial^* E_1 \cup \partial^* E_2 - \partial^* E) = 0$$

and hence for any open set A:

$$\int_A |D\varphi_{E_1}| + \int_A |D\varphi_{E_2}| = H_{n-1}(\partial^* E_1 \cap A) + H_{n-1}(\partial^* E_2 \cap A)$$

$$\leqq H_{n-1}(\partial^* E \cap A) = \int_A |D\varphi_E|.$$

Comparing with (15.1) we obtain (15.2). Suppose now that E has least boundary in A, and let F be a set coinciding with E_1 outside some compact set $K \subset A$. We have

$$\int_A |D\varphi_F| + \int_A |D\varphi_{E_2}| \geqq \int_A |D\varphi_{F \cup E_2}| \geqq \int_A |D\varphi_E|$$

since $F \cup E_2$ coincides with $E = E_1 \cup E_2$ outside K. The minimality of E_1 now follows from (15.2). In a similar way we can prove that E_2 has least perimeter. $\square$

15.4 Lemma: *Let Ω be an open set in $\mathbb{R}^n$, and let $v : \Omega \to \mathbb{R}$ be a solution of the minimal surface equation. Let V be the subgraph of v and let $E \subset Q = \Omega \times \mathbb{R}$ be a Caccioppoli set coinciding with V outside some compact set $K \subset Q$. Then*

$$(15.3) \quad \int_K |D\varphi_V| \leq \int_K |D\varphi_E|$$

with equality holding if and only if $E = V$.

Proof: Inequality (15.3) is an immediate consequence of Theorem 14.8. The last assertion follows from the uniqueness of the solution to the minimal surface equation (Proposition 12.10).   □

15.5   Theorem: Let E be a minimal cone in $\mathbb{R}^n$ with vertex at 0, and suppose that E is contained in the half-space $H = \{x \in \mathbb{R}^n : x_n < 0\}$. Then $E = H$.

Proof: The theorem is trivially true if $n = 2$. Suppose that it holds for $n - 1$. We show first that it is possible to assume that $\partial E \cap \partial H$ contains only singular points for ∂E. Actually, if $x_0 \in \partial^* E \cap \partial H$, then ∂H is the tangent plane to ∂E at x_0 and in a neighborhood of x_0, ∂E can be represented as the graph of a non-negative solution u of the minimal surface equation. From the strong maximum principle (Theorem C.7) we have $u \equiv 0$ near x_0. Since the singular set of ∂E, $\partial E - \partial^* E$, has dimension less than $n - 7$, we can conclude that $\partial E \supset \partial H$ and therefore $E = H - E_1$, with $\partial H \cap \partial E_1 \subset \partial E - \partial^* E$ and therefore $H_{n-1}(\bar{H}' \cap \bar{E}_1) = 0$, $H' = \mathbb{R}^n - H$. From Lemma 15.3 applied to $E' = \mathbb{R}^n - E$ it follows that E_1 is a minimal cone and satisfies $\partial^* E_1 \cap \partial H = \emptyset$.

We now split into cases as follows

(a) $\partial E \cap \partial H = \{0\}$. This means that for every $\varepsilon > 0$ the set $E_\varepsilon = E - H_\varepsilon = E - \{x \in \mathbb{R}^n : x_n < -\varepsilon\}$ is relatively compact and non empty. Since both H_ε and, by Remark 15.2, $H_\varepsilon \cup E = H_\varepsilon \cup E_\varepsilon$ have least perimeter, if A is an open set containing $\bar{E}_\varepsilon$ we have

$$\int_A |D\varphi_{H_\varepsilon}| = \int_A |D\varphi_{H_\varepsilon - E_\varepsilon}|$$

contradicting the preceding Lemma.

(b) $\partial E \cap \partial H \ni x_0 \neq 0$. In this case the half-line Ox_0 is contained in $\partial E \cap \partial H$. If we blow up E near x_0 (taking $E_j = \{x \in \mathbb{R}^n : x_0 + j^{-1}x \in E\}$ and letting $j \to \infty$) we get a minimal cylinder C satisfying the hypothesis of the theorem, and therefore a minimal cone with vertex in 0 contained in a half-space in $\mathbb{R}^{n-1}$. Since we have supposed that the result holds in $\mathbb{R}^{n-1}$, this cone must coincide with the half-space and thus the theorem is proved.   □

Before proceeding further, we recall some definitions.

Let A be an open set in $\mathbb{R}^n$, and let $v \in BV(A)$. We say that v is a supersolution if for every non-negative $\varphi \in BV(A)$ with compact support we have

$$\int_A \sqrt{1 + |Dv|^2} \leq \int_A \sqrt{1 + |D(v + \varphi)|^2}.$$

When $v \in W^{1,1}(A)$, it is easily seen that v is a supersolution if and only if

$$\int \frac{D_i v D_i \varphi}{\sqrt{1 + |Dv|^2}} dx \geq 0 \qquad \forall \varphi \in C_0^1(A), \; \varphi \geq 0.$$

Moreover, if $v \in C^2(A)$, this is equivalent to

$$(15.4) \quad D_i \left\{ \frac{D_i v}{\sqrt{1 + |Dv|^2}} \right\} \leq 0.$$

Let v be a supersolution in A, and let V be its subgraph. It follows easily from Theorem 14.8 that for any compact set $K \subset C \subset\subset A \times \mathbb{R}$ we have

$$(15.5) \quad \int_C |D\varphi_V| \leq \int_C |D\varphi_{V \cup K}|.$$

Now let Ω be an open set with C^2-boundary $\partial\Omega$, and let $x_0 \in \partial\Omega$. If $\partial\Omega$ has non-negative mean curvature it is possible to represent it near x_0 as the graph of a function v satisfying (15.4). This means that there exists an open set B containing x_0 such that

$$\int_B |D\varphi_\Omega| \leq \int_B |D\varphi_{\Omega \cup K}|$$

for any compact set $K \subset B$.

15.6   Definition: Let Ω be a Caccioppoli set, and let B be an open set with $B \cap \partial\Omega \neq \emptyset$. We say that $\partial\Omega$ has non-negative mean curvature in B if

$$(15.6) \quad \int_B |D\varphi_\Omega| \leq \int_B |D\varphi_{\Omega \cup K}|$$

for every compact set $K \subset B$. If $x_0 \in \partial\Omega$ and $\partial\Omega$ has non-negative mean curvature in some open set B containing x_0 we say that the mean curvature of $\partial\Omega$ is non-negative near x_0.

15.7   Lemma: *Let $\partial\Omega$ have non-negative mean curvature near x_0, let $Q = \Omega \times \mathbb{R}$ and let $t_0 \in \mathbb{R}$. Then ∂Q has non-negative mean curvature near $z_0 = (x_0, t_0)$.*

Proof: Let B be a neighborhood of x_0 such that (15.6) holds, and let $C = B \times (t_0 - 1, t_0 + 1)$. If $K \subset C$ is compact and $|t - t_0| < 1$, we set

$$K_t = \{x \in \mathbb{R}^n : (x, t) \in K\}.$$

We have

$$\int_B |D\varphi_\Omega| \leqq \int_B |D\varphi_{\Omega \cup K_t}|$$

and the conclusion follows by integrating over t and by remarking that by Lemma 9.8

$$\int_{t_0 - 1}^{t_0 + 1} dt \int_B |D\varphi_{\Omega \cup K_t}| \leqq \int_C |D\varphi_{\Omega \cup K}| \qquad \qquad \square$$

From Lemma 15.1 we get easily

15.8   Proposition: *Let B be a ball in $\mathbb{R}^{n+1}$ and let $U \subset Q$ be a Caccioppoli set with least perimeter in $\overline{Q \cap B}$. Suppose that ∂Q has non-negative mean curvature in B. Then U minimizes the perimeter in B.*

Proof: We may suppose that $P(U) < +\infty$. Let E be any set coinciding with U outside some compact set contained in B. Since U minimizes the perimeter in $\overline{Q \cap B}$ we have

$$\int_B |D\varphi_U| \leqq \int_B |D\varphi_{Q \cap E}|.$$

On the other hand we have

$$\int_B |D\varphi_Q| \leqq \int_B |D\varphi_{Q \cup E}|$$

and therefore from (15.1)

$$\int_B |D\varphi_{E\cap Q}| \leqq \int_B |D\varphi_E|.$$

In conclusion

$$\int_B |D\varphi_U| \leqq \int_B |D\varphi_E|. \qquad \qquad \Box$$

We can now state our boundary regularity theorem.

15.9   Theorem [MM7]: *Let Ω be a bounded open set in $\mathbb{R}^n$ with Lipschitz-continuous boundary $\partial\Omega$, and let u be a minimum for the functional*

$$\mathscr{I}(u) = \int_\Omega \sqrt{1 + |Du|^2} + \int_{\partial\Omega} |u - \varphi| dH_{n-1}.$$

Suppose that $\partial\Omega$ has non-negative mean curvature near x_0 and that φ is continuous at x_0. Then

(15.7) $\lim\limits_{x \to x_0} u(x) = \varphi(x_0).$

Proof: If we had

$$\lim\limits_{x \to x_0} \sup u(x) > \varphi(x_0)$$

there would exist a sequence $x_j \to x_0$ such that

(15.8) $\lim\limits_{j \to \infty} u(x_j) = \lambda > \varphi(x_0).$

We shall prove that this would lead to a contradiction.

Assume first that $\partial\Omega$ is of class C^1 near x_0, and suppose that (15.8) holds. Since φ is continuous at x_0, there exists a ball $B = B(z_0, R)$, centred at $z_0 = (x_0, \lambda)$, not intersecting the graph of φ. This implies that U minimizes the perimeter in $\overline{Q \cap B}$. Since ∂Q has non-negative mean curvature near z_0, we can conclude from Proposition 15.8 that ∂U has least area in B. We now blow up U near z_0 obtaining a sequence

$$U_j = \{z \in \mathbb{R}^{n+1}; j^{-1}z + z_0 \in U\}$$

of minimal sets in $B(0, jR)$, with

$$U_j \subset Q_j = \{z \in \mathbb{R}^{n+1}; j^{-1}z + z_0 \in Q\}.$$

Arguing as in Theorem 9.2 we conclude that a subsequence of U_j converges to a minimal cone C with vertex at 0. Since $\partial\Omega$ is C^1 near x_0 the sequence Q_j converges to a half-space containing C.

From Theorem 15.5 it follows that C is a half-space, and from Theorem 9.4 we conclude that ∂U is regular near z_0. This means that near z_0, both ∂U and ∂Q can be represented as graphs of two functions w and v, respectively a solution and a supersolution of the minimal surface equation, with $w \leq v$ and $w = v$ at some interior point. From the strong maximum principle it follows that $w \equiv v$ and therefore $\partial U \equiv \partial Q$ near z_0, a conclusion that contradicts (15.8). We have therefore

$$\lim_{x \to x_0} \sup u(x) \leq \varphi(x_0)$$

and since $-u$ minimizes the area with boundary datum $-\varphi$:

$$\lim_{x \to x_0} \inf u(x) \geq \varphi(x_0)$$

so that (15.7) holds. In order to conclude the proof of the theorem we shall show that if (15.8) is satisfied, then $\partial\Omega$ must be smooth near x_0. For that, suppose that ∂Q can be written as the graph of a Lipschitz-continuous function $x_1 = w(x_2, \ldots, x_{n+1})$ in a ball $\mathscr{B} \subset \mathbb{R}^n$ centred at $\xi = (x_2^0, \ldots, x_n^0, \lambda)$, with $x_1 < w(x_2, \ldots, x_{n+1})$ in Q.

Let v be a function minimizing the area functional with boundary datum w on $\partial\mathscr{B}$. Since w is a supersolution we have $v \leq w$ in $\mathscr{B}$. We shall prove that $v(\xi) = w(\xi)$. Suppose on the contrary that $v(\xi) < w(\xi)$, and for $\varepsilon > 0$ let U^ε be the set U translated by $-\varepsilon$ in the x_1-direction:

$$U^\varepsilon = \{x \in \mathbb{R}^{n+1} : (x_1 + \varepsilon, x_2, \ldots, x_{n+1}) \in U\}.$$

Since $(w(\xi), \xi) \in U$, for ε small enough the set $U^\varepsilon - V$ is compact and non-empty. By Remark 15.2 it follows that $V \cap U^\varepsilon$ has least perimeter, contradicting Lemma 15.4. We have therefore $v(\xi) = w(\xi)$. From the strong maximum principle (Theorem C7) we conclude that $v = w$ in B, so that w is regular there. $\square$

In particular, if $\partial\Omega$ has non-negative mean curvature and φ is continuous, we have $u \in C(\bar{\Omega})$ and $u = \varphi$ on $\partial\Omega$. Hence Theorem 13.6 is a special case of Theorem 15.9. When the mean curvature of $\partial\Omega$ is negative, we cannot say in general that $u = \varphi$. Nevertheless L. Simon has proved some interesting regularity results which we will only state, referring to the original papers for the proofs.

15.10   Theorem [SL1]: *Let $\partial\Omega$ be of class C^4 in a neighborhood B of x_0, and let $\partial\Omega \cap B$ have strictly negative mean curvature. Suppose that φ is Lipschitz-continuous on $\partial\Omega \cap B$, and let u minimize the functional $\mathscr{J}$. Then u is Hölder continuous in $\bar{\Omega} \cap B$ and the trace of u is Lipschitz-continuous in $\partial\Omega \cap B$.*

Simon has also investigated in [SL3] the behaviour of u at those points $x_0 \in \partial\Omega$ where the mean curvature $\mathscr{H}$ of $\partial\Omega$ is zero. In general, u may have discontinuities at these points, even if $\mathscr{H} \leq 0$ near x_0 and φ is constant. On the other hand if $\mathscr{H}$ changes sign at $x_0 \in \partial\Omega$ and if $\delta\mathscr{H}(x_0) \neq 0$ (δ denoting the tangential gradient on $\partial\Omega$) then u is continuous at x_0. A more general result is proved in [WG].

We shall conclude this section with a brief discussion of the uniqueness of the function minimizing the area functional

$$\mathscr{J}(u) = \int_\Omega \sqrt{1 + |Du|^2} + \int_{\partial\Omega} |u - \varphi| dH_{n-1}.$$

We have already seen that two minima of $\mathscr{J}$ may at most be different by an additive constant (Proposition 14.12). From Theorem 15.9 we deduce immediately that if Ω is bounded, $\partial\Omega$ is of class C^2 and φ is continuous, then $\mathscr{J}$ has exactly one minimum point. To show that, we only have to notice that, since Ω is bounded and $\partial\Omega$ is smooth, there is at least one point $x_0 \in \partial\Omega$ where the mean curvature of $\partial\Omega$ is positive (actually one can show more, namely that all the principal curvatures of $\partial\Omega$ are positive at some point $x_0 \in \partial\Omega$).

Since φ is continuous, we have

$$\lim_{x \to x_0} u(x) = \varphi(x_0)$$

for any minimizing function u, and the uniqueness follows from the above-mentioned Proposition 14.12.

The assumption that $\partial\Omega$ is smooth can be released. In fact, we have

15.11   Theorem: *Let Ω be a bounded open set in $\mathbb{R}^n$ with Lipschitz-continuous boundary and let φ be a continuous function on $\partial\Omega$. Then the functional $\mathscr{J}$ has exactly one minimum point.*

Proof: Let $\mathcal{B}$ be a ball containing Ω with $\partial\mathcal{B} \cap \partial\Omega = \{x_0\}$, and let $\varepsilon > 0$. Since φ is continuous, there exists $a \in \mathbb{R}^n$ such that for every $x \in \partial\Omega$:

$$\varphi(x) \leqq \varphi(x_0) + \varepsilon + \langle a, x - x_0 \rangle$$

The maximum principle now implies that $u(x) \leqq \varphi(x_0) + \varepsilon + \langle a, x - x_0 \rangle$ in Ω and therefore

$$\limsup_{x \to x_0} u(x) \leqq \varphi(x_0) + \varepsilon.$$

In a similar way:

$$\liminf_{x \to x_0} u(x) \geqq \varphi(x_0) - \varepsilon$$

and hence

$$\lim_{x \to x_0} u(x) = \varphi(x_0)$$

from which uniqueness follows. $\square$

We remark that the point x_0 here need not be a point of positive mean curvature in the sense of Definition 15.6. When φ is not continuous the above arguments fail, and we may have non-uniqueness.

15.12   Example [SE]: Let Ω be the set in figure, bounded by four similar arcs of a circumference of radius ρ centred at the points $(\pm 1, \pm 1)$. Taking $1 < \rho < \sqrt{2}$, the boundary of Ω is Lipschitz-continuous.

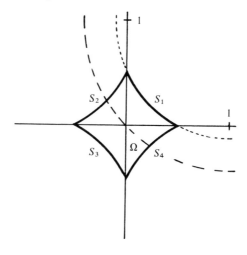

Let $\varphi = M$ on S_1 and S_3, $\varphi = -M$ on S_2 and S_4, where M is a constant that we shall fix later. Due to the symmetry of the data and to the convexity of $\mathcal{J}$, we may conclude that if $z(x, y)$ minimizes $\mathcal{J}$, the same is true for

$$v(x, y) = \tfrac{1}{2}(z(x, y) - z(-x, y))$$

and

$$u(x, y) = \tfrac{1}{2}(v(x, y) - v(x, -y)).$$

In conclusion we can suppose that our minimum $u(x, y)$ satisfies

(15.9) $u(x, 0) = u(0, y) = 0.$

Consider now the annulus of interior radius ρ and exterior radius $\sqrt{2}$ centred at (1.1), and let $w(x, y)$ be the solution of the minimal surface equation constructed in Example 12.15. We have

$$w = 0 \text{ on } \partial B_{\sqrt{2}}; \qquad 0 \leq w \leq M_0$$

$$\frac{\partial w}{\partial r} = +\infty \text{ on } \partial B_\rho.$$

Let us consider the set Ω_1, intersection of Ω with the first quadrant. We have $0 = u(x, 0) \leq w(x, 0)$; $0 = u(0, y) \leq w(0, y)$ and since $\dfrac{\partial w}{\partial v} = -\infty$ on S_1 we may apply the maximum principle (Lemma 12.13) concluding that $u \leq w$ in Ω_1. In particular, taking $M > M_0$ we have $u \leq M_0 < \varphi$ on S_1. The same argument shows that $u \leq M_0 < \varphi$ on S_3, whereas $u \geq -M_0 > \varphi$ on $S_2 \cup S_4$. It is now a simple matter of computation to show that if $|c| < M - M_0$ we have $\mathcal{J}(u + c) = \mathcal{J}(u)$. $\square$

16. A Further Extension of the Notion of Non-Parametric Minimal Surface

The theory developed in the two previous sections, and in particular the notion of non-parametric surfaces as graphs of BV functions, though very useful in many respects, is not completely free from limitations.

In the first place, the functions under examination belong to $BV(\Omega)$, and therefore must have finite area. This precludes the treatment of the Dirichlet problem in unbounded domains Ω of infinite measure, or even in bounded domains with infinite data.

A second unpleasant feature, which contrasts strongly with the parametric case, is that – generally speaking – limits of non-parametric minimal surfaces may not be non-parametric in the above sense (i.e. graphs of BV-functions). A simple example of such behaviour is a sequence of hyperplanes converging to a vertical hyperplane. In general, to conclude that the limit is again the graph of a BV-function, one has to get uniform estimates (e.g. for the BV-norm, or even for the L^1-norm, on compact subdomains) that even when they exist are not always simple to prove.

In order to avoid such disadvantages, we shall extend the notion of non-parametric minimal surface in such a way that these new objects share the simplicity of the graphs with the flexibility of the general parametric surfaces.

Our starting point will be the relation between a function $u \in BV_{\text{loc}}(\Omega)$ and its subgraph

$$U = \{(x, t) \in \Omega \times \mathbb{R}; \, t < u(x)\}.$$

We have proved in 14.6 that for any open set $A \subset\subset \Omega$ we have

$$\int_A \sqrt{1 + |Du|^2} = \int_{A \times \mathbb{R}} |D\varphi_U|.$$

Moreover, u is a local minimum for the area in Ω if and only if U has locally least perimeter in $\Omega \times \mathbb{R}$ (Theorem 14.9).

These remarks justify the following definition. We shall consider functions $u : \Omega \to [-\infty, +\infty]$, taking possibly the values $+\infty$ and $-\infty$ on sets of positive measure, and their corresponding subgraphs U.

16.1 Definition: Let $u:\Omega \rightarrow [-\infty, +\infty]$ be a measurable function. We say that u is a *quasi-solution* of the minimal surface equation (briefly: a quasi-solution) in Ω if its subgraph U locally minimizes the perimeter in $Q = \Omega \times \mathbb{R}$.

In other words, u is a quasi-solution if for every $V \subset Q$, coinciding with U outside some compact set $K \subset Q$, we have

$$\int\limits_K |D\varphi_U| \leq \int\limits_K |D\varphi_V|$$

It follows from Theorem 14.9 that every non-parametric minimal surface is a quasi-solution. In particular every classical solution is a quasi-solution. Conversely, if a quasi-solution belongs to $L^1_{loc}(\Omega)$, then it is locally bounded in Ω (Theorem 14.10) and therefore it is a non-parametric minimal surface. Moreover, u is smooth in Ω and it is a solution of the minimal surface equation.

16.2 Example: Let $E \subset \Omega$ have least perimeter in Ω. Then

$$(16.1) \quad u(x) = \begin{cases} +\infty & x \in E \\ -\infty & x \notin E \end{cases}$$

is a quasi-solution.

Let V coincide with $U = E \times \mathbb{R}$ outside some compact set $K \subset Q = \Omega \times \mathbb{R}$. Let $A \subset\subset \Omega$ and $T > 0$ be such that $K \subset A_T = A \times (-T, T)$.

For $-T < t < T$ set

$$V_t = \{x \in \Omega : (x, t) \in V\}.$$

We have $V_t = E$ outside A and hence

$$\int\limits_A |D\varphi_E| \leq \int\limits_A |D\varphi_{V_t}|.$$

Integrating with respect to t and using Lemma 9.8:

$$\int\limits_{A_T} |D\varphi_U| = 2T \int\limits_A |D\varphi_E| \leq \int\limits_{-T}^{T} dt \int\limits_A |D\varphi_{V_t}| \leq \int\limits_{A_T} |D\varphi_V|.$$

The converse is also true; namely, if u in (16.1) is a quasi-solution then E has least area in Ω.

For otherwise there would exist a compact set $K \subset \Omega$, a positive number δ, and a set F coinciding with E outside K such that

$$\int_K |D\varphi_F| \leq \int_K |D\varphi_E| - \delta$$

For $T > 0$ set

$$F_T = \begin{cases} F \times \mathbb{R} & \text{in } K_T = K \times [-T, T] \\ E \times \mathbb{R} = U & \text{outside } K_T \end{cases}$$

We have

$$\int_{K_T} |D\varphi_{F_T}| \leq \int_{K_T} |D\varphi_{F \times \mathbb{R}}| + 2|K| \leq \int_{K_T} |D\varphi_U| - 2T\delta + 2|K| <$$

$$< \int_{K_T} |D\varphi_U|$$

whenever $T\delta > |K|$. But in this case U would not be a local minimum for the area in $\Omega \times \mathbb{R}$.

16.3   Lemma: *Let $\{u_k\}$ be a sequence of measurable functions in Ω, and suppose that the characteristic functions of the subgraphs, φ_{U_k}, converge in $L^1_{\text{loc}}(Q)$ to φ_U. Then U is the subgraph of a measurable function $u:\Omega \to [-\infty, +\infty]$ and a subsequence extracted from $\{u_k\}$ converges almost everywhere to u.*

Proof: For $x \in \Omega$ and $V \subset Q$ we define

$$V^x = \{t \in \mathbb{R} : (x, t) \in V\}.$$

For every compact set $K \subset \Omega$ and every $T > 0$ we have

$$\lim_{k \to \infty} \int_K dx \int_{-T}^{T} |\varphi_{U_k^x} - \varphi_{U^x}| \, dt = 0$$

and therefore passing to a subsequence we can suppose that

$$\lim_{k \to \infty} \int_{-T}^{T} |\varphi_{U_k^x} - \varphi_{U^x}| \, dt = 0$$

for every $T > 0$ and for almost every $x \in \Omega$.

Since U_k^x is the half-line $(-\infty, u_k(x))$, the set U^x must be itself a half-line (possibly $\varnothing$ or $\mathbb{R}$) for almost each x. Setting

$$u(x) = \sup U^x$$

we get immediately that U is the subgraph of u, and that $u_k(x)$ converge almost everywhere to $u(x)$.   □

We note that the converse is also trivially true; namely that if $u_k \to u$ a.e. in Ω then $\varphi_{U_k} \to \varphi_U$ in $L_{loc}^1(Q)$.

We can now prove the following result, which gives a negative answer to the problem of regularity of parametric minimal surfaces.

16.4   Theorem [BDGG]: *The Simons cone*

$$S = \{x \in \mathbb{R}^{2n} : x_1^2 + \ldots + x_n^2 > x_{n+1}^2 + \ldots + x_{2n}^2\}$$

has least perimeter in $\mathbb{R}^{2n}$, for $n \geq 4$.

Proof [MAM]: For $x \in \mathbb{R}^{2n}$ we set

$$y = (x_1, x_2, \ldots, x_n); \ z = (x_{n+1}, \ldots, x_{2n})$$

and

$$v(x) = (|y|^2 - |z|^2)(|y|^2 + |z|^2).$$

If $\mathscr{E}$ is the minimal surface operator defined in (12.9) we find easily:

$$\mathscr{E}(v) = 4(|y|^2 - |z|^2)\{(n + 2)[1 + 16(|y|^6 + |z|^6)] -$$

$$- 48(|y|^2 + |z|^2)(|y|^4 + |z|^4)\}$$

For $n \geq 4$ the quantity within brackets is positive and therefore v is a subsolution ($\mathscr{E}(v) \geq 0$) in S and a supersolution in $\mathbb{R}^{2n} - \bar{S}$. If we set for $k > 0$,

$$v_k(x) = k^{-1}v(kx) = k^3 v(x)$$

we can prove easily that

$$\mathscr{E}(v_k)(x) = k\mathscr{E}(v)(kx)$$

and hence the same conclusion holds for v_k.

Now consider the Dirichlet problem for the area functional (12.1) in the ball B_R, with boundary datum v_k on ∂B_R, and let u_k be its unique solution (Theorem 12.10). From the symmetry of v_k we deduce that $u_k = 0$ on ∂S and therefore from the maximum principle (Lemma 12.13) we conclude that $u_k \geq v_k$ in S and $u_k \leq v_k$ in $B_R - \bar{S}$.

Let now $k \to \infty$; the sequence v_k (and hence u_k) tends to $+\infty$ in S and to $-\infty$ in $B_R - \bar{S}$. By Lemma 16.3 the function

$$u(x) = \begin{cases} +\infty & x \in S \\ -\infty & x \in B_R - \bar{S} \end{cases}$$

is a quasi-solution in B_R, so that from Example 16.2 we conclude that S minimizes perimeter in B_R. $\qquad\square$

A second consequence of Lemma 16.3 is the following general compactness result for quasi-solutions.

16.5   Proposition: *Every sequence $\{u_k\}$ of quasi-solutions in Ω has a subsequence converging almost everywhere to a quasi-solution.*

Proof: Let K be a compact set contained in $Q = \Omega \times \mathbb{R}$. Without loss of generality we can suppose that K has smooth boundary. Comparing U_j with $U_j - K$ we get at once

$$\int_K |D\varphi_{U_j}| \leq H_n(\partial K)$$

and therefore by Theorem 1.19 we can extract from φ_{U_j} a subsequence converging in $L^1(K)$. By means of a diagonal procedure we can select a subsequence U_k such that φ_{U_k} converges in $L^1_{\mathrm{loc}}(Q)$ to the subgraph U of some function u, and such that $u_k \to u$ almost everywhere in Ω.

By Lemma 9.1, the set U has least perimeter in Q, and hence u is a quasi-solution. $\qquad\square$

As we have remarked, a quasi-solution u may take the values $+\infty$ and $-\infty$ on sets of positive measure. We set

$$(16.2) \quad P = \{x \in \Omega : u(x) = +\infty\}$$

and

$$(16.3) \quad N = \{x \in \Omega : u(x) = -\infty\}.$$

In the following we shall investigate the circumstances under which a quasi-solution is a classical solution.

A first step in this direction consists in the study of the properties of P and N.

We remark that if u is a quasi-solution, the same is true for $-u$, and that a change of sign in u interchanges P and N. It is therefore sufficient to discuss the properties of one of these sets, for instance of P.

16.6   Theorem: Let u be a quasi-solution in $Q = \Omega \times \mathbb{R}$. Then P has locally least perimeter in Ω.

Proof: The functions

$$u_j(x) = u(x) - j$$

are obviously quasi-solutions in Q. As $j \to \infty$, the sequence u_j converges almost everywhere to the quasi-solution

$$u(x) = \begin{cases} +\infty & x \in P \\ -\infty & x \notin P \end{cases}$$

and therefore as in the Example 16.2 we conclude that P minimizes the area. □

Before proceeding further, we recall that we can always assume that for every set E under consideration the inequalities

$$0 < |E \cap \mathscr{B}_R(x)| < \omega_n R^n$$

hold for each $R > 0$ and $x \in \partial E$ (see Remark 3.2).

From Proposition 5.14 and from the minimality of P we infer at once that for every $x_0 \in P$ and every R, $0 < R < \text{dist}(x_0, \partial \Omega)$ we have:

$$|P \cap \mathscr{B}_R(x_0)| \geq cR^n.$$

This implies that if $A \subset \Omega$ is open and $|P \cap A| = 0$, then $P \cap A = \varnothing$.

Even better, if $P \neq \varnothing$ then

$$|P| > c\delta^n$$

where $\delta = \sup_{x \in P} \text{dist}(x, \partial\Omega)$.

16.7 Proposition: *Let u be a quasi-solution in Ω, and let $P = \emptyset$. Then u is locally bounded above in Ω.*

Proof: Suppose not. Then there exists a sequence $\{x_j\}$, convergent to a point $x_0 \in \Omega$ such that $u(x_j) > j$.

Let $2R < \text{dist}(x_0, \partial\Omega)$, and suppose that $|x_j - x_0| < R$.

Let U_j be the subgraph of $u_j(x) = u(x) - j$. We have $u_j(x_j) > 0$ and therefore the points $z_j = (x_j, 0)$ belong to U_j.

By Proposition 5.14 we have

$$|U_j \cap B_R(z_j)| > cR^{n+1}$$

and hence

$$(16.4) \quad |U_j \cap B_{2R}(z_0)| > cR^{n+1}.$$

When $j \to \infty$, U_j converge locally to $P \times \mathbb{R}$, which would be non-empty by (16.4). □

In particular, if both P and N are empty (or, equivalently, if P and N have zero measure) the quasi-solution u is locally bounded and therefore it is a classical solution.

This remark and Proposition 16.5 show the flexibility of the notion of quasi-solution. Given a sequence of quasi-solutions (in particular of solutions) one can always select a subsequence converging to a quasi-solution, without any a-priori estimate.

The limit will be a classical solution if P and N are empty. In this case it is sometimes possible to get a-posteriori uniform bounds for the norms of the converging solutions.

As a first example of the use of the method we shall consider the problem of the removal of singularities.

16.8 Lemma [MM9]: *Let Q be an open set in $\mathbb{R}^{n+1}$, and let S be a closed set in Q, with $H_n(S) = 0$. Let U be a set of least perimeter in $Q - S$. Then U has least perimeter in Q.*

Proof: Since $H_n(S) = 0$, there exists a sequence $\{S_j\}$ of open sets with piecewise smooth boundary such that $S_j \supset S_{j+1} \supset \cdots \supset S$ and

$$(16.5) \quad |S_j| \to 0, \quad H_n(\partial S_j) \to 0.$$

Now let V be a set coinciding with U outside some compact set $K \subset Q$. Setting

$$V_j = \begin{cases} V & \text{in } Q - S_j \\ U & \text{in } S_j \end{cases}$$

we have $V_j = U$ outside $K - S_j \subset Q - S$ and therefore if A is any open set with $K \subset A \subset\subset \Omega$:

$$\int_{A - S_j} |D\varphi_U| \leqq \int_{A - S_j} |D\varphi_{V_j}| \leqq \int_A |D\varphi_V| + H_n(\partial S_j).$$

Passing to the limit as $j \to \infty$ we get

$$(16.6) \quad \int_{A - S} |D\varphi_U| \leqq \int_A |D\varphi_V|.$$

On the other hand we have from Theorem 4.4:

$$\int_A |D\varphi_U| = H_n(\partial^* U \cap A) = H_n(\partial^* U \cap (A - S)) = \int_{A - S} |D\varphi_U|$$

since $H_n(S) = 0$.
Comparing the last equation with (16.6) we get the conclusion of the lemma. $\qquad\qquad\square$

16.9   Theorem (*Removal of singularities*): *Let Ω be an open set in $\mathbb{R}^n$, and let Σ be a closed set in Ω with $H_{n-1}(\Sigma) = 0$.*
Let $u \in C^2(\Omega - \Sigma)$ be a solution of the minimal surface equation

$$(16.7) \quad \sum_{i=1}^n \frac{\partial}{\partial x_i} \left\{ \frac{\partial u/\partial x_i}{\sqrt{1 + |Du|^2}} \right\} = 0$$

in $\Omega - \Sigma$. Then u can be extended to a solution of (16.7) in the whole of Ω.

Proof: Let U be the subgraph of u; U has least perimeter in $Q - S$ ($Q = \Omega \times \mathbb{R}$, $S = \Sigma \times \mathbb{R}$).

Since S is closed, and $H_n(S) = 0$, U minimizes the perimeter in Q, so that u is a quasi-solution in Ω.

On the other hand, since $u \in C^2(\Omega - \Sigma)$, we must have $P \subset \Sigma$ and $N \subset \Sigma$ and hence, since Σ has zero measure, $P = N = \varnothing$.

It follows from Proposition 16.7 that u is locally bounded in Ω, and hence u can be extended to a classical solution in Ω. $\square$

The above theorem was proved by J. C. C. Nitsche [NI2] in the case $n = 2$, and by E. De Giorgi and G. Stampacchia [DGS] with the additional assumption that Σ is compact (so that the singular set does not touch $\partial\Omega$. We note that this is practically satisfied in dimension $n = 2$ due to the condition $H_1(\Sigma) = 0$). The proof given here is due to M. Miranda [MM9]. Different proofs have been obtained independently at about the same time by G. Anzellotti [AN] and L. Simon [SL2].

The reader will recognize the similarity between Theorem 16.8 and analogous results for harmonic functions, and more generally for solutions of elliptic partial differential equations. We remark that here we do not need any assumption on the behavior of u near Σ.

We shall turn now to quasi-solution of the Dirichlet problem, in situations that cannot be treated by means of the theory developed in section 14, as for instance the Dirichlet problem with infinite data. As we shall see, this problem fits quite well in the framework of quasi-solutions.

The simplest example of non-parametric minimal surface taking infinite values at the boundary is the so-called Scherk's surface:

$$u(x, y) = \log \cos x - \log \cos y$$

in the square $\Omega = \{(x, y) \in \mathbb{R}^2 : |x| < \pi/2, |y| < \pi/2\}$.

The existence of non-parametric minimal surfaces taking the values $+\infty$ or $-\infty$ on prescribed parts of the boundary, and possibly finite values on the other portions of $\partial\Omega$, was discussed by Jenkins and Serrin [JS1] in the·two-dimensional case. Later, J. Spruck [SP] extended this result to surfaces of constant mean curvature, and U. Massari [MU] proved the existence for general mean curvature and arbitrary dimension. We shall follow here his proof.

Let us begin with the definition of quasi-solution for the Dirichlet problem with (possibly) infinite data. Let Ω be a bounded open set in $\mathbb{R}^n$ with Lipschitz continuous boundary $\partial\Omega$, and let $\mathscr{B}$ be a ball containing $\bar{\Omega}$. Let $\psi : \mathscr{B} \to [-\infty, +\infty]$ be a measurable function such that its subgraph Ψ is a Caccioppoli set (i.e. has locally finite perimeter) in $\mathscr{B} \times \mathbb{R}$.

16.10   Definition: A measurable function $u : \mathscr{B} \to [-\infty, +\infty]$ is a quasi-solution of the Dirichlet problem with boundary datum ψ on $\partial\Omega$ if:

i) $U = \Psi$ outside $\Omega \times \mathbb{R}$
ii) for any set V coinciding with U outside some compact set $K \subset \bar{\Omega} \times \mathbb{R}$
    we have

$$\int_K |D\varphi_U| \leq \int_K |D\varphi_V|.$$

We have the following existence result.

16.11   Theorem: *Let Ω be a bounded domain with Lipschitz-continuous boundary, and let ψ be a measurable function in whose subgraph Ψ is a Caccioppoli set in $\mathscr{B} \times \mathbb{R}$.*

The Dirichlet problem with boundary datum ψ on $\partial\Omega$ admits a quasi-solution.

Proof: For $k > 0$ we set

$$\psi_k(x) = \begin{cases} k & \text{if } \psi(x) > k \\ \psi(x) & \text{if } |\psi(x)| \leq k \\ -k & \text{if } \psi(x) < -k. \end{cases}$$

The function ψ_k belongs to $BV(\mathscr{B})$, and therefore the Dirichlet problem with data ψ_k on the boundary has a solution $u_k \in BV(\mathscr{B})$ (Theorem 14.5).

In order to go to the limit as $k \to +\infty$ we want to derive a uniform estimate for the perimeter of U_k over compact subset of $\mathscr{B} \times \mathbb{R}$.

For $T > 0$ and $A \subset \mathscr{B}$ let $A_T = A \times (-T, T)$, and let $W = U_k \cup \Omega_T$. We have

$$\int_{\mathscr{B}_{2T}} |D\varphi_{U_k}| \leq \int_{\mathscr{B}_{2T}} |D\varphi_W| \leq \int_{\mathscr{B}_{2T}} |D\varphi_{\Psi_k}| + 4TH_{n-1}(\partial\Omega) + 2|\Omega| \leq c(T)$$

and therefore by Theorem 1.19, arguing as in Lemma 16.3 we can extract a subsequence, that we shall denote again by u_k, converging almost everywhere to a measurable function u.

Let U be the subgraph of u. It is clear that $U = \Psi$ outside $\Omega \times \mathbb{R}$. Now let V be a Caccioppoli set in $\mathscr{B} \times \mathbb{R}$, coinciding with U outside some compact set $K \subset \bar{\Omega} \times \mathbb{R}$. Let $T > 0$, let A be an open set, $\Omega \subset\subset A \subset\subset \mathscr{B}$, such that $K \subset A_T$, and define

$$V_k = \begin{cases} V & \text{in } A_T \\ U_k & \text{outside } A_T. \end{cases}$$

If $k > T$, V_k coincides with U_k outside $\bar{\Omega}_T \subset A_{2T}$, and therefore:

$$\int\limits_{A_{2T}} |D\varphi_{U_k}| \leqq \int\limits_{A_{2T}} |D\varphi_{V_k}|.$$

This inequality is equivalent to

$$(16.8) \quad \int\limits_{A_T} |D\varphi_{U_k}| \leqq \int\limits_{A_T} |D\varphi_V| + \int\limits_{\partial A_T} |\varphi_{U_k} - \varphi_U| dH_n$$

since $V = U$ outside $K \subset A_T$.

We can now choose T in such a way that the last integral in (16.8) tends to zero when $k \to \infty$. Passing to the limit and using the lower semi-continuity of the perimeter (Theorem 1.9) we get the conclusion of the theorem. $\square$

In general the quasi-solution u given by Theorem 16.10 takes the values $+\infty$ or $-\infty$ in nonempty subsets of Ω. Our goal is to give necessary and sufficient conditions under which the quasi-solution u is a classical solution in Ω, and takes the value ψ on $\partial\Omega$. We shall begin by deriving necessary conditions.

16.12 General Hypothesis: In order to simplify the discussion we shall make some additional assumptions on Ω and ψ. More precisely, we shall suppose that $\partial\Omega$ is Lipschitz-continuous and that there exist three open sets A_0, A_+ and A_- with the properties:

  i) $\psi(x) = \pm\infty$ in $A_\pm$
  ii) ψ is continuous in A_0
  iii) $\partial\Omega = \Gamma_0 \cup \Gamma_+ \cup \Gamma_- \cup \mathscr{N}$, where $\Gamma_0 = \partial\Omega \cap A_0$, $\Gamma_\pm = \partial\Omega \cap A_\pm$ and $H_{n-1}(\mathscr{N}) = 0$.
  iv) Γ_+ and Γ_- are C^2-hypersurfaces.

The next result improves Theorem 16.5.

16.13 Theorem: *Let u be a quasi-solution of the Dirichlet problem with boundary data ψ on $\partial\Omega$. Then P minimizes the perimeter among all sets coinciding with A_+ outside Ω.*

Proof: With the argument of Theorem 16.5 we prove that the function

$$u_\infty(x) = \begin{cases} +\infty & x \in P \\ -\infty & \text{otherwise} \end{cases}$$

is a quasi-solution with boundary value

$$\psi_\infty(x) = \begin{cases} +\infty & x \in A_+ \\ -\infty & \text{otherwise.} \end{cases}$$

The conclusion follows as in Example 16.2.   □

It is easily seen that the conclusion of the preceding theorem is equivalent to the assertion that $P \subset \Omega$ minimizes the functional

$$(16.9) \quad \mathscr{I}_+(E) = \int_\Omega |D\varphi_E| + \int_{\partial\Omega} |\varphi_E - \varphi_{\Gamma_+}| dH_{n-1}.$$

Similarly, N minimizes the perimeter among all sets coinciding with A_- outside Ω, and hence $N \subset \Omega$ minimizes

$$(16.10) \quad \mathscr{I}_-(E) = \int_\Omega |D\varphi_E| + \int_{\partial\Omega} |\varphi_E - \varphi_{\Gamma_-}| dH_{n-1}$$

among all sets $E \subset \Omega$.

If we want u to be a classical solution in Ω (i.e. $P \cap \Omega = N \cap \Omega = \emptyset$), a necessary condition is therefore that the empty set is a minimum for $\mathscr{I}_+$ and $\mathscr{I}_-$.

In particular, this implies that Γ_+ and Γ_- have non-positive mean curvature, (see 15.6).

A second necessary condition is given by the following.

16.14 Proposition: *Let $x_0 \in \partial\Omega$ and suppose that there exists a solution $u(x)$ of the minimal surface equation in Ω, such that*

$$(16.11) \quad \lim_{x \to x_0} u(x) = +\infty.$$

Let $\partial\Omega$ be smooth near x_0. Then the mean curvature of $\partial\Omega$ at x_0 is non negative.

Proof: Suppose on the contrary that the mean curvature at x_0 is less than zero. Then we can construct an open set $\Omega_1 \subset \Omega$, with C^2-boundary, such that $\partial\Omega_1 \cap \partial\Omega = \{x_0\}$ and with negative mean curvature at x_0.

For $x \in \Omega_1$ let $d_1(x) = \text{dist}(x, \partial\Omega_1)$ and let $R > 0$ be such that

$$\Delta d_1 \geq \varepsilon_0 > 0 \quad \text{in } \Omega_1 \cap \mathcal{B}_R(x_0).$$

As in 12.14, the function

$$v(x) = \alpha - \beta d_1^{1/2}, \; \beta > \varepsilon_0^{-1/2}$$

is a supersolution in $\Omega_1 \cap \mathcal{B}_R$. Taking

$$\alpha = \sup_{\partial\mathcal{B}_R \cap \Omega_1} u + \beta R^{1/2}$$

we have $v \geq u$ on $\partial\mathcal{B}_R \cap \Omega_1$ and hence by Lemma 12.13 $v \geq u$ in $\mathcal{B}_R \cap \Omega_1$. This contradicts (16.11). $\qquad\square$

Combining the two preceding results we obtain the necessary condition for the existence of a classical solution that Γ_+ and Γ_- have zero mean curvature.

With regard to Γ_0, we recall that a necessary and sufficient condition for the general solvability of the Dirichlet problem with continuous boundary data was that the mean curvature of $\partial\Omega$ be non-negative (see section 12). This condition is also necessary in the more general case under examination.

16.15   Proposition: *Let u be a classical solution of the minimal surface equation in Ω taking the value ψ on $\partial\Omega$, and let $x_0 \in \Gamma^+$. Then*

$$(16.12) \; \lim_{x \to x_0} Tu(x) = v(x_0)$$

where v is the exterior normal to $\partial\Omega$ and

$$Tu = \frac{Du}{\sqrt{1 + |Du|^2}}$$

Proof: Let A_+ be as in 16.12, and let $\mathcal{B}_R \subset A_+$ be a ball centred at x_0 such that $\partial\Omega \cap \mathcal{B}_R \subset \Gamma_+$ and $\partial\Omega \cap \mathcal{B}_R$ is the graph of some function w. Since Γ_+ has zero mean curvature, w is a solution of the minimal surface equation and therefore $A_+ - \Omega$ has least perimeter in $\mathcal{B}_R$ (Theorem 14.9). Arguing as in Example 16.2 we infer that $L = (A_+ - \Omega) \times \mathbb{R}$ has boundary of least area in $\mathcal{B}_R \times \mathbb{R}$.

Now let $x_j \to x_0$, and let $u_j(x) = u(x) - u(x_j)$.

The subgraph U_j of u_j minimizes perimeter in $\overline{\Omega} \times \mathbb{R}$ and therefore by Proposition 15.7 in $\mathscr{B}_R \times \mathbb{R}$.

When $j \to \infty$ we have $u(x_j) \to +\infty$ and hence U_j converges to L in $L^1_{\mathrm{loc}}(\mathscr{B}_R \times \mathbb{R})$. By Theorem 9.4 we get

$$v^L(x_0, 0) = \lim_{j \to \infty} v^{U_j}(x_j, 0).$$

On the other hand we have $v^L(x_0, 0) = (-v(x_0), 0)$ and

$$v^{U_j}(x_j, 0) = (-Tu(x_j), (1 + |Du(x_j)|^2)^{-1/2})$$

proving (16.12). □

In a similar way we show that $\lim_{x \to x_0} Tu(x) = -v(x_0)$ for $x_0 \in \Gamma_-$. As an immediate corollary we get

16.16 Theorem (*Uniqueness of the solution*): *Let Ω and ψ be as in 16.12, and consider the Dirichlet problem*

$$(16.13) \quad \begin{cases} \operatorname{div} Tu = 0 & in\ \Omega \\ u = \psi & on\ \partial\Omega. \end{cases}$$

Then:

  i) *if $\Gamma_0 \neq \varnothing$ the problem has at most one classical solution;*
  ii) *if $\Gamma_0 = \varnothing$ any two solutions of (16.13) differ by an additive constant.*

Proof: It follows at once from (16.12) and Lemma 12.13. □

Let us continue the investigation of necessary conditions for the existence of solutions to the Dirichlet problem (16.13).

16.17 Proposition: *Suppose that there exists a vector field v in Ω such that*

  i) $|v| < 1$ *in Ω*
  ii) $\langle v, v \rangle = 1$ *on Γ_+ (v is the exterior normal to $\partial\Omega$)*
  iii) $\operatorname{div} v = 0$ *in Ω.*

Then if $E \subset \Omega$, $E \neq \emptyset$, Ω we have

(16.14) $\mathscr{J}_{+}(\emptyset) < \mathscr{J}_{+}(E)$

where $\mathscr{J}_{+}$ is the functional defined in (16.9).

Proof: Integrating iii) over E we get

$$0 = \int_{E} \operatorname{div} v \, dx = \int_{\partial^{*}E} \langle v, v^{E} \rangle \, dH_{n-1}$$

where $\partial^{*}E$ is the reduced boundary of E, and v^{E} is the exterior normal to $\partial^{*}E$ (see section 3). Now

(16.15) $\int_{\partial^{*}E} \langle v, v^{E} \rangle \, dH_{n-1} = \int_{\partial\Omega} \varphi_{E} \langle v, v \rangle \, dH_{n-1} + \int_{\Omega \cap \partial^{*}E} \langle v, v^{E} \rangle \, dH_{n-1}$

and therefore from i) and ii):

(16.16) $-\int_{\Gamma_{+}} \varphi_{E} dH_{n-1} + \int_{\partial\Omega - \Gamma_{+}} \varphi_{E} dH_{n-1} + \int_{\Omega} |D\varphi_{E}| > 0.$

The conclusion follows at once by adding to both sides of (16.6) the quantity $\mathscr{J}_{+}(\emptyset) = H_{n-1}(\Gamma_{+})$. $\qquad\qquad\qquad\qquad\qquad\qquad\qquad\qquad\qquad\qquad\square$

As usual, a symmetric result holds for the functional $\mathscr{J}_{-}$ given by (16.10), if we replace condition ii) with

$\qquad\qquad$ ii') $\langle v, v \rangle = -1 \quad$ on Γ_{-}.

Taking $v = Tu$ we obtain at once the following necessary condition.

16.18 **Proposition:** *Suppose that the Dirichlet problem (16.13) has a classical solution. Then for any set $E \subset \Omega$, we have*

(16.17) $\mathscr{J}_{\pm}(\emptyset) \leq \mathscr{J}_{\pm}(E),$

with equality holding at most for $E = \emptyset$ and $E = \Omega$. $\qquad\qquad\qquad\qquad\square$

The validity of the strict inequality in (16.17) for $E = \Omega$ depends on the set Γ_{0}. It is easily seen that when $\Gamma_{0} = \emptyset$ and $E = \Omega$ we must have the equality in (16.17), since in this case $\mathscr{J}_{\pm}(\emptyset) = \mathscr{J}_{\pm}(\Omega)$.

On the other hand, when Γ_0 is non-empty and has non-negative mean curvature we have the strict inequality even for $E = \Omega$.

Suppose on the contrary that $\mathscr{J}_+(\varnothing) = \mathscr{J}_+(\Omega)$.

From (16.15) with $E = \Omega$ and $v = Tu$ we get

$$H_{n-1}(\Gamma_+) + \int_{\partial\Omega - \Gamma_+} \langle v, Tu \rangle \, dH_{n-1} = 0$$

and therefore from $\mathscr{J}_+(\varnothing) = \mathscr{J}_+(\Omega)$ we get $\langle v, Tu \rangle = -1$ on $\partial\Omega - \Gamma_+$ and in particular on Γ_0.

We will show that this leads to a contradiction. Let $\mathscr{B}_R$ be a ball centred on Γ_0 and such that $\partial\Omega \cap \mathscr{B}_R \subset \Gamma_0$. The boundary of $\Omega_R = \Omega \cap \mathscr{B}_R$ has non-negative mean curvature, and therefore the Dirichlet problem with continuous data on $\partial\Omega_R$ is generally solvable. On the other hand we have a solution $u(x)$ of the minimal surface equation in Ω_R such that $\langle v, Tu \rangle = -1$ on $\partial\Omega \cap \mathscr{B}_R$. By Lemma 12.13 we deduce that any solution v of div $Tv = 0$ in Ω_R, with $v \geq u$ on $\partial\mathscr{B}_R \cap \Omega$ satisfies $v \geq u$ on Ω_R, and therefore cannot take boundary values less than u on $\partial\Omega \cap \mathscr{B}_R$. This contradicts the general solvability of the Dirichlet problem.

We have thus proved

16.19   Theorem: *Let Ω and ψ satisfy assumptions 16.12, and let $\Gamma_0 \neq \varnothing$. Then necessary conditions for the general solvability of the Dirichlet problem*

$$(16.18) \quad \begin{cases} \operatorname{div} Tu = 0 & \text{in } \Omega \\ u = \psi & \text{on } \partial\Omega - \mathscr{N} \end{cases}$$

are the following:

(16.19) $\partial\Omega$ has non-negative mean curvature,

(16.20) the empty set is the only minimum of the functionals $\mathscr{J}_+$ and $\mathscr{J}_-$.

If instead we have $\Gamma_0 = \varnothing$, condition (16.20) has to be replaced with

(16.20)′ $\varnothing$ and Ω are the only minima for $\mathscr{J}_+$ and $\mathscr{J}_-$. $\square$

We shall show now that the above conditions are sufficient for the existence of a solution to the problem (16.18). Let us begin with the simpler case $\Gamma_0 \neq \varnothing$.

16.20   Theorem: *Let* Ω *and* ψ *satisfy assumptions* 16.12; *let* $\Gamma_0 \neq \emptyset$ *and suppose that* (16.19) *and* (16.20) *hold. Let* u *be a quasi-solution of the Dirichlet problem with boundary datum* ψ. *Then* u *is a classical solution of the minimal surface equation in* Ω *and satisfies* $u(x) = \psi(x)$ *on* $\partial\Omega - \mathcal{N}$.

Proof: We have only to show that $u = \psi$ on $\partial\Omega - \mathcal{N}$. Let us begin with a point $x_0 \in \Gamma_0$. Suppose by contradiction that there exists a sequence $x_k \to x_0$ such that $u(x_k) \to l > \psi(x_0)$. We have two possibilities:

(I) $l < +\infty$. This case can be excluded exactly as in Theorem 15.8.

(II) $l = +\infty$. Let $\mathcal{B}_R$ be a ball centred at x_0 such that $\partial\Omega \cap \mathcal{B}_R$ is the graph of a function and is contained in Γ_0. Arguing as in Theorem 15.8 we conclude that the subgraph U of u has least perimeter in $\mathcal{B}_R \times (M, +\infty)$, where $M = \sup\limits_{\mathcal{B}_R} \psi + 1$.

In particular if $u(x_k) > M + R$ we get from Proposition 5.14 the estimate

$$|U \cap B_{R/2}(z_k)| > cR^{n+1}$$

where $z_k = (x_k, u(x_k))$ and $c > 0$. Since $U = \Psi$ outside $\Omega \times \mathbb{R}$ we have $U \cap B_{R/2}(z_k) \subset \Omega \times \mathbb{R}$, and therefore for every $T > 0$:

$$\text{meas}\{(x, t); x \in \Omega, T < t < u(x)\} \geq cR^{n+1}.$$

But then we should have $P \cap \Omega \neq \emptyset$, a contradiction.

The same argument works for $x_0 \in \Gamma_-$, and therefore we can conclude that

$$(16.21) \quad \limsup_{x \to x_0} u(x) \leq \psi(x_0)$$

for every $x_0 \in \partial\Omega - \mathcal{N}$.

Changing ψ into $-\psi$ (and hence u into $-u$) we obtain

$$\liminf_{x \to x_0} u(x) \geq \psi(x_0)$$

and comparing with (16.21):

$$\lim_{x \to x_0} u(x) = \psi(x_0). \qquad \qquad \square$$

The treatment of the case $\Gamma_0 = \emptyset$ is slightly more complex, since we cannot exclude in principle the possibilities $P = \Omega$ or $N = \Omega$.

Actually, $u \equiv +\infty$ and $u \equiv -\infty$ are quasi-solutions in this case, and the simple argument of Theorem 16.11 can lead exactly to one of these. We therefore have to choose carefully the sequence ψ_k in order to avoid divergent solutions.

We assume of course that necessary conditions (16.19) and (16.20)′ hold.

We start from the remark that the Dirichlet problem with data

$$\psi_\alpha(x) = \begin{cases} +\infty & \text{on } \Gamma_+ \\ \alpha & \text{on } \Gamma_- \end{cases}$$

cannot have a solution, since otherwise the functional $\mathcal{J}_+$ would have the empty set as the unique minimum (Theorem 16.19) contradicting the necessary condition (16.20)′. We can therefore conclude that the only quasi-solution to the Dirichlet problem with data ψ_α is $u \equiv +\infty$.

Similarly, $u \equiv -\infty$ is the only quasi-solution for the Dirichlet problem with data

$$\psi^\alpha(x) = \begin{cases} \alpha & \text{on } \Gamma_+ \\ -\infty & \text{on } \Gamma_-. \end{cases}$$

Let x_0 a point in Ω that will remain fixed, and consider the Dirichlet problem with boundary data

$$\psi_\beta^\alpha(x) = \begin{cases} \alpha > 0 & \text{on } \Gamma_+ \\ \beta < 0 & \text{on } \Gamma_-. \end{cases}$$

This problem has a unique solution (see Proposition 14.12 and Theorem 15.8), which we shall denote by v_β^α. We shall prove that it is possible to choose α and β arbitrarily large in such a way that $v_\beta^\alpha(x_0) = 0$.

Let $k \in \mathbb{N}$, and suppose that $v_{-k}^k(x_0) > 0$. When $\beta \to -\infty$, $v_\beta^k(x_0)$ decreases, and tends to $-\infty$. Moreover $v_\beta^k(x_0)$ is a continuous function of β (this can be seen by combining the uniqueness theorem 15.11 with the a priori estimate for the gradient 13.5).

It is therefore possible to find a value $\beta(k) < -k$ such that $v_{\beta(k)}^k(x_0) = 0$. In a similar way, if $v_{-k}^k(x_0) < 0$, we find a value $\alpha(k) > k$ such that $v_{-k}^{\alpha(k)}(x_0) = 0$.

Let us define $u_k = v_{\beta(k)}^k$ or $u_k = v_{-k}^{\alpha(k)}$ in the two cases. To the sequence u_k we can apply the argument of theorem 16.10, from which we conclude the existence of a quasi-solution u to the Dirichlet problem with infinite data.

Condition (16.10)′ says that in order to show that u is a classical solution, we have only to exclude the possibilities $u \equiv +\infty$ or $u \equiv -\infty$; in other words we have to show that neither P nor N can coincide with Ω.

This follows from the fact that $u_k(x_0) = 0$. In fact, let U_k be the subgraph of

u_k, and let $R < \text{dist}(x_0, \partial\Omega)$. The point $z_0 = (x_0, 0)$ belongs to ∂U_k, and therefore from Proposition 5.14 we get

$$|U_k \cap B_R(z_0)| \geqq cR^{n+1}$$

and

$$|B_R(z_0) - U_k| \geqq cR^{n+1}.$$

These inequalities imply at once that U and $\Omega \times \mathbb{R} - U$ must have positive measure in $B_R(z_0)$ and therefore that P and N must both be empty.

The quasi-solution u is therefore a classical solution in Ω; repeating the argument of Theorem 16.19 we show that $u = \psi$ on $\partial\Omega - \mathcal{N}$.

We have thus proved

16.21   Theorem: *Let Ω and ψ satisfy assumptions 16.12; let $\Gamma_0 = \varnothing$ and suppose the necessary conditions (16.19) and (16.20)' hold.*

Then there exists a function $u(x)$ satisfying in Ω the minimal surface equation

$$\text{div } Tu = 0$$

and such that

$$\lim_{x \to x_0} u(x) = \psi(x_0)$$

for every $x_0 \in \partial\Omega - \mathcal{N}$.

17. The Bernstein Problem

In 1915, N. S. Bernstein proved his celebrated theorem concerning entire minimal graphs [BS2]:

17.1 Theorem: *Let u(x, y) be a solution of the minimal surface equation in the plane:*

$$(17.1) \quad \frac{\partial}{\partial x}\left(\frac{u_x}{\sqrt{1 + u_x^2 + u_y^2}}\right) + \frac{\partial}{\partial y}\left(\frac{u_y}{\sqrt{1 + u_x^2 + u_y^2}}\right) = 0.$$

Then the graph of u is a plane.

It is worth remarking the similarity between Bernstein's and Liouville's theorem. As in the case of the removal of singularities, the main difference consists in the fact that Bernstein's theorem does not require any additional assumption on u.

We may write the equation of minimal surfaces in the form

$$(17.2) \quad (1 + u_y^2)u_{xx} - 2u_x u_y u_{xy} + (1 + u_x^2)u_{yy} = 0.$$

We remark now that (17.2) is the necessary and sufficient condition for the existence of a function $\Phi(x, y)$ such that

$$\Phi_{xx} = \frac{1 + u_x^2}{\sqrt{1 + u_x^2 + u_y^2}}$$

$$\Phi_{xy} = \frac{u_x u_y}{\sqrt{1 + u_x^2 + u_y^2}}$$

$$\Phi_{yy} = \frac{1 + u_y^2}{\sqrt{1 + u_x^2 + u_y^2}}.$$

Such a function Φ satisfies the equation

$$(17.3) \quad \Phi_{xx}\Phi_{yy} - \Phi_{xy}^2 = 1.$$

We have the following result.

17.2   Theorem: *If Φ satisfies (17.3) in the plane, it is a polynomial of degree two.*

Proof: [NI1]. Changing possibly its sign, we can suppose that Φ is convex, and therefore the map

$$\xi = x + \Phi_x(x, y)$$

$$\eta = y + \Phi_y(x, y)$$

is a diffeomorphism of $\mathbb{R}^2$ onto itself.
   If we set $\zeta = \xi + i\eta$ and

$$w(\zeta) = x - \Phi_x(x, y) - i(y - \Phi_y(x, y))$$

it is easily seen that w is an entire holomorphic function. Moreover, we have

$$|w'(\zeta)|^2 = \frac{\Phi_{xx} + \Phi_{yy} - 2}{\Phi_{xx} + \Phi_{yy} + 2} < 1.$$

By Liouville's theorem, w' is constant. On the other hand

$$\Phi_{xx} = \frac{|1 - w'|^2}{1 - |w'|^2} = c_1$$

$$\Phi_{yy} = \frac{|1 + w'|^2}{1 - |w'|^2} = c_2$$

and therefore Φ is a polynomial of degree two.
   Coming back to our function u we see that u_x and u_y are constant, and therefore u is an affine function, thus proving the Bernstein theorem.   □
   It might be worth giving another short proof of theorem 17.1. For that, let U be the subgraph of u. Since U has least perimeter in $\mathbb{R}^3$ we have from section 10.8:

$$\int_{\partial U} \{|\delta\zeta|^2 - c^2\zeta^2\} dH_2 \geqq 0$$

for every Lipschitz-continuous function ζ with compact support in $\mathbb{R}^3$. For $j \in \mathbb{N}$, set

$$\zeta_j(x) = \begin{cases} 1 & |x| \leq j \\ 0 & |x| \geq j^2 \\ 2 - \dfrac{\log|x|}{\log j} & j < |x| < j^2. \end{cases}$$

We have for every j:

(17.4) $\displaystyle \int_{\partial U \cap B_j} c^2 \, dH_2 \leq \int_{\partial U} |\delta \zeta_j|^2 \, dH_2 \leq \int_{\partial U} |D\zeta_j|^2 \, dH_2.$

Let now $\Sigma_j = \partial U \cap (B_{j^2} - B_j)$; we have

$$\int_{\partial U} |D\zeta_j|^2 \, dH_2 = \frac{1}{(\log j)^2} \int_{\Sigma_j} |x|^{-2} \, dH_2$$

$$= \frac{1}{(\log j)^2} \int_0^\infty H_2\{x \in \Sigma_j : |x|^{-2} > t\} \, dt.$$

The last integral is not greater than

$$\int_{j^{-4}}^{j^{-2}} H_2(\partial U \cap B_{t^{-1/2}}) \, dt + j^{-4} H_2(\Sigma_j)$$

and since $H_2(\partial U \cap B_r) \leq Ar^2$ we have

$$\int_{\partial U} |D\zeta_j|^2 \, dH_2 \leq A/\log j.$$

Letting $j \to \infty$ we obtain from (17.4) $c^2 \equiv 0$ and therefore ∂U is a plane.   □

It is natural to ask whether Bernstein's theorem is true in higher dimension; i.e. whether the only solutions of the minimal surface equation in $\mathbb{R}^n$ are polynomials of the first degree. This is what we shall call the Bernstein problem.

It is easily seen that none of the above techniques can be extended to dimension $n > 2$.

Actually, the first one is based on complex variable methods, and the second one relies essentially on the fact that in two variables any bounded set has zero absolute capacity; or more precisely that

$$\inf\{\int |Df|^2 dx; f \in C_0^1(\mathbb{R}^2), f \geq 1 \text{ on } E\} = 0$$

for every bounded set $E \subset \mathbb{R}^2$. It is well known that this infimum is positive in $\mathbb{R}^n$, $n > 2$.

A new idea suitable for extension was provided by Fleming [FW] and is embodied in the following theorem.

17.3   Theorem: *Let U be a set with least perimeter in $\mathbb{R}^n$. Then either $n \geq 8$ or ∂U is a hyperplane.*

Proof: For $j \in \mathbb{N}$ set

$$U_j = \{x \in \mathbb{R}^n : jx \in U\}.$$

The sets U_j have least perimeter in $\mathbb{R}^n$.

Arguing as in Theorem 9.2 we conclude that a subsequence U_{r_j} converges to a minimal cone C. Moreover, we have for almost every $R > 0$

$$(17.5) \quad \omega_{n+1} \leq (Rr_j)^{1-n} \int\limits_{B_{Rr_j}} |D\varphi_U| = R^{1-n} \int\limits_{B_R} |D\varphi_{U_{r_j}}| \to R^{1-n} \int\limits_{B_R} |D\varphi_C|.$$

Suppose now that $n \leq 7$. In this case the cone C must be a half-space, and therefore

$$R^{1-n} \int\limits_{B_R} |D\varphi_C| = \omega_{n-1}.$$

Using the monotonicity of $R^{1-n} \int\limits_{B_R} |D\varphi_U|$ we find from (17.5)

$$R^{1-n} \int\limits_{B_R} |D\varphi_U| = \omega_{n-1}$$

for every $R > 0$. It follows from (5.11) that U is a cone. But then $U = C$ and ∂U is a hyperplane. $\square$

A similar argument gives a weaker form of the Bernstein's theorem for minimal sets. The similarity with Liouville's theorem is evident.

17.4   Theorem: *Let U be a set of least perimeter in $\mathbb{R}^n$. If U contains a half-space, ∂U is a hyperplane.*

Proof: Arguing as above we find a minimal cone C with vertex at 0. Since C contains a half-space we conclude by Theorem 15.5 that C is a half-space. The conclusion now follows at once.   □

The conclusion of Theorem 17.3 is the best possible, since in dimension $n = 8$ we have the Simons cone

$$S = \{x \in \mathbb{R}^8 : x_1^2 + \ldots + x_4^2 > x_5^2 + \ldots + x_8^2\}.$$

which is minimal in $\mathbb{R}^8$ (Theorem 16.4).

However, when U is the subgraph of a function, we can say something more. For that we will need some auxiliary results, that are of interest by themselves and represent weaker forms of Bernstein's theorem.

17.5   Theorem [MJ]: *Let u be a solution of the minimal surface equation*

$$(17.6) \quad \sum_{i=1}^{n} D_i \left(\frac{D_i u}{\sqrt{1 + |Du|^2}} \right) = 0$$

in $\mathbb{R}^n$. Suppose further that u has bounded gradient in $\mathbb{R}^n$. Then u is an affine function.

Proof: The function $w = \partial u / \partial x_s$ ($s = 1, \ldots, n$) satisfies the equation

$$(17.7) \quad \frac{\partial}{\partial x_i} \left(a_{ij}(x) \frac{\partial w}{\partial x_j} \right) = 0$$

with

$$a_{ij}(x) = \frac{\delta_{ij}(1 + |Du|^2) - D_i u D_j u}{(1 + |Du|^2)^{3/2}} \in L^\infty(\mathbb{R}^n)$$

(see section 12). Since $|Du|$ is bounded in $\mathbb{R}^n$ we have for every $\xi \in \mathbb{R}^n$

$$\nu |\xi|^2 \leq a_{ij}(x) \xi_i \xi_j$$

with some constant $\nu > 0$, and therefore equation (17.7) is uniformly elliptic. Since w is bounded, the non-negative function $z = w - \inf w$ satisfies the same equation (17.7), and therefore Harnack's inequality (Theorem C.2)

$$\sup_{\mathscr{B}_R} z \leqq c \inf_{\mathscr{B}_R} z$$

holds for every $R > 0$, with c independent of R. Letting $R \to +\infty$ we get $\sup_{\mathbb{R}^n} z = 0$ and therefore $w = $ constant.   □

As a consequence of the a priori estimate for the gradient (Theorem 13.5) we have

17.6   Theorem: *Let u be a solution in $\mathbb{R}^n$ of the minimal surface equation. Suppose that for every $x \in \mathbb{R}^n$:*

(17.8) $u(x) \leqq K(1 + |x|)$

for some constant K.
   Then u is an affine function.

Proof: Let $x_0 \in \mathbb{R}^n$. We have

$$\sup_{\mathscr{B}_R(x_0)} u \leqq K(1 + R + |x_0|)$$

and therefore from (13.11)

$$|Du(x_0)| \leqq \exp\{c_5(1 + K) + c_5 R^{-1}(|u(x_0)| + K(1 + |x_0|))\}.$$

Letting $R \to +\infty$ we get

$$|Du(x_0)| \leqq \exp c_5(1 + K) = c_6$$

and the conclusion follows from the preceding theorem.   □
   It is clear that, using (13.12) instead of (13.11) we can replace (17.8) with

$$u(x) \geqq -K(1 + |x|).$$

In particular, the conclusion of the theorem holds if we suppose $u(x) \geqq a + \langle b, x \rangle$, in close analogy with Liouville's theorem for harmonic functions.

17.7   Lemma: *Let Ω be an open set in $\mathbb{R}^n$, and let $\{u_j\}$ be a sequence of quasi-solutions in Ω (see Definition 16.1), converging a.e. to a quasi-solution v. Suppose that*

$$P = \{x \in \Omega : v(x) = +\infty\} = \varnothing.$$

Then for every compact $K \subset \Omega$ there exists a constant $c(K)$ such that

$$\sup_j \sup_{x \in K} u_j(x) \leq c(K).$$

Proof: By Proposition 16.7, the function v is locally bounded above. For $K \subset \Omega$ let $2d = \operatorname{dist}(K, \partial\Omega)$ (if $\Omega = \mathbb{R}^n$ set $d = 1$) and let

$$\gamma = \gamma(K) = \sup_{x \in K_d} v(x)$$

$$(K_d = \{x \in \mathbb{R}^n : \operatorname{dist}(x, K) \leq d\}).$$

We shall prove that

(17.9) $\limsup_{j \to \infty} \sup_{x \in K} u_j(x) \leq \gamma(K).$

Otherwise, there would exist ε, $0 < \varepsilon < d$, a subsequence u_j^* and a sequence $x_j \in K$ such that

$$z_j = (x_j, \gamma(K) + \varepsilon) \in U_j^*.$$

From Proposition 5.11 we obtain

$$|U_j^* \cap B(z_j, \varepsilon)| \geq \alpha(n)\varepsilon^{n+1}$$

and therefore

(17.10) $|U_j^* \cap K_\varepsilon \times (\gamma, \gamma + 2\varepsilon)| \geq \alpha\varepsilon^{n+1}.$

Since the u_j's converge a.e. to v (and therefore $\varphi_{U_j} \to \varphi_V$) we have from (17.10):

$$|V \cap K_\varepsilon \times (\gamma, \gamma + 2\varepsilon)| \geq \alpha\varepsilon^{n+1} > 0$$

whereas from the definition of γ

$$V \cap K_\varepsilon \times (\gamma, \gamma + 2\varepsilon) = \emptyset.$$ □

We can now prove the extension of Bernstein's theorem.

17.8   Theorem: *Let $u: \mathbb{R}^n \to \mathbb{R}$ be an entire solution of the minimal surface equation*

$$\sum_{i=1}^{n} D_i \left\{ \frac{D_i u}{\sqrt{1 - |Du|^2}} \right\} = 0.$$

Then either $n \geq 8$ or the graph of u is a hyperplane.

Proof: Let U be the subgraph of u, and let U_j be as in Theorem 17.3. The set U_j is the subgraph of the function

$$u_j(x) = \frac{1}{j} u(jx).$$

Arguing as in Theorem 17.3 we construct a minimal cone C in $\mathbb{R}^{n+1}$ as the limit of a subsequence U_{r_j}. The cone C is itself a subgraph of a quasi-solution v.

Let

$$P = \{ x \in \mathbb{R}^n : v(x) = +\infty \}$$

$$N = \{ x \in \mathbb{R}^n : v(x) = -\infty \}.$$

(I) $P = \emptyset$. In this case v is locally bounded above in $\mathbb{R}^n$. By Lemma 17.7 the functions u_{r_j} are equibounded in the unit ball B_1, and hence

$$\sup_{\mathscr{B}_{r_j}} u(x) \leq \gamma r_j.$$

From the a-priori estimate of the gradient (13.11) we have

$$\sup_{S_{r_{j/6}}} |Du| \leq \exp \left\{ c_5 \left(1 + \gamma - \frac{u(x_0)}{r_j} \right) \right\}.$$

Letting $j \to +\infty$ we obtain $\sup |Du| \leq \text{const}$ and therefore by Theorem 17.5 u is an affine function. The same conclusion holds if $N = \varnothing$.

(II)   We shall prove now that case (I) must occur if $n \leq 7$. Suppose on the contrary that P and N are both non-empty. Since C is a cone, P and N are minimal cones in $\mathbb{R}^n$, with vertex in 0. Since $n \leq 7$, P and N are half-space and therefore

$$v(x) = \begin{cases} +\infty & x \in P \\ -\infty & x \in N = \mathbb{R}^n - P. \end{cases}$$

This means that C is a half-space and ∂C is a vertical hyperplane. Arguing as in Theorem 17.3 we conclude that $U = C$. This is impossible because ∂U cannot be a vertical hyperplane. $\square$

17.9   Remark: We note that the same proof works if only one supposes that u is a quasisolution. It is easily seen that this result cannot be ameliorated since if S in the Simons cone the function

$$f(x) = \begin{cases} +\infty & x \in S \\ -\infty & x \notin S \end{cases}$$

is a quasisolution in $\mathbb{R}^8$.

The construction of a similar example with a true solution f requires some additional work.

In the first place, let us introduce some notation. We set

$$u = (x_1^2 + x_2^2 + x_3^2 + x_4^2)^{1/2}$$

$$v = (x_5^2 + x_6^2 + x_7^2 + x_8^2)^{1/2}$$

and

(17.11)
$$\begin{cases} T = \{(u,v) \in \mathbb{R}^2 : u \geq 0,\ v \geq 0\} \\[2mm] T_1 = \{(u,v) \in \mathbb{R}^2 : 0 \leq v < u\} \\[2mm] T_2 = \{(u,v) \in \mathbb{R}^2 : 0 \leq u < v\} \end{cases}$$

We shall consider functions depending on x only through u and v:

$$f(x) = F(u, v).$$

For such functions, the minimal surface equation (17.6) (or rather (12.9)) becomes

$$(17.12)\ \mathscr{E}(F) = (1 + F_v^2)F_{uu} - 2F_uF_vF_{uv} + (1 + F_u^2)F_{vv} +$$

$$+ 3\left(\frac{F_u}{u} + \frac{F_v}{v}\right)(1 + F_u^2 + F_v^2) = 0$$

Now let

$$(17.13)\ F_1(u, v) = (u^2 - v^2)(u^2 + v^2)^{1/2}$$

and

$$(17.14)\ F_2(u, v) = H\left\{(u^2 - v^2)\left[1 + (u^2 + v^2)^{1/2}\left(1 + A\left|\frac{u^2 - v^2}{u^2 + v^2}\right|^{\lambda - 1}\right)\right]\right\}$$

where $\frac{21}{16} < \lambda < \frac{4}{3}$, H is defined by

$$H(z) = \int_0^z \exp\left(B\int_{|w|}^{+\infty} \frac{dt}{t^{2-\lambda}(1 + t^{3(\lambda - 1)})}\right) dw$$

and A and B are constants to be determined. We shall prove later that

$$(17.15)\ \begin{array}{ll} \mathscr{E}(F_1) > 0 & \text{in } \mathring{T}_1 \\ \mathscr{E}(F_1) < 0 & \text{in } \mathring{T}_2 \end{array}$$

whereas, if A and B are sufficiently large:

$$(17.16)\ \begin{array}{ll} \mathscr{E}(F_2) < 0 & \text{in } \mathring{T}_1 \\ \mathscr{E}(F_2) > 0 & \text{in } \mathring{T}_2 \end{array}$$

Moreover we have

$$0 < F_1 < F_2 \qquad \text{in } T_1$$

$$F_2 < F_1 < 0 \qquad \text{in } T_2$$

$$F_1 = F_2 = 0 \qquad \text{if } u = v.$$

Setting as usual

$$S = \{x \in \mathbb{R}^8 : x_1^2 + \cdots + x_4^2 > x_5^2 + \cdots + x_8^2\}$$

and $f_i(x) = F_i(u, v)(i = 1, 2)$ we have in S

$$\mathscr{E}(f_1) > 0$$
$$\mathscr{E}(f_2) < 0$$

and

$$0 < f_1 < f_2.$$

Similarly, in $\mathbb{R}^8 - \bar{S}$:

$$\mathscr{E}(f_1) < 0$$
$$\mathscr{E}(f_2) > 0$$

and

$$f_2 < f_1 < 0.$$

Moreover,

$$f_1 = f_2 = 0 \qquad \text{in } \partial S.$$

Now let $R > 0$ and let $f^{(R)}$ be the solution of the Dirichlet problem:

$$\begin{cases} \mathscr{E}(f) = 0 & \text{in } \mathscr{B}_R \\ f = f_1 & \text{on } \partial \mathscr{B}_R. \end{cases}$$

Since f_1 is smooth on $\partial \mathscr{B}_R$, the existence and uniqueness of the solution are guaranteed by Theorem 12.10. In particular from the uniqueness of the solution it follows that

$$f^{(R)}(x) = F^{(R)}(u, v)$$

for some function $F^{(R)}$. Moreover, since $F_1(u, v) = -F_1(v, u)$ the same is true for $F^{(R)}$ and therefore

$$f^{(R)}(x) = 0 \qquad \text{on } \partial S.$$

By the maximum principle 12.5 we have

$$0 \leq f_1(x) \leq f^{(R)}(x) \leq f_2(x) \qquad \text{in } S \cap \mathcal{B}_R$$

and similarly

$$f_2(x) \leq f^{(R)}(x) \leq f_1(x) \leq 0 \qquad \text{in } \mathcal{B}_R - S,$$

whence

$$(17.17) \; |f_1(x)| \leq |f^{(R)}(x)| \leq |f_2(x)| \qquad \text{in } \mathcal{B}_R.$$

Consider now the sequence $f^{(h)}(h = 1, 2, \ldots)$ and let $\rho > 0$. We have from the a-priori estimate of the gradient (Theorem 13.5)

$$\sup_{x \in \mathcal{B}_\rho} |Df^{(h)}(x)| \leq \exp\left\{ c_5\left(1 + \sup_{\mathcal{B}_{\rho+1}} |f^{(h)}(x)| \right) \right\}$$

and from (17.17):

$$(17.18) \; \sup_{x \in \mathcal{B}_\rho} |Df^{(h)}(x)| \leq \exp\left\{ c_5\left(1 + \sup_{\mathcal{B}_{\rho+1}} |f_2(x)| \right) \right\} \leq c_6(\rho)$$

for every $h > \rho + 1$.

We note that the right-hand side of (17.18) is independent of h. By the Ascoli-Arzelà theorem, it is possible to extract from $\{f^{(h)}\}$ a subsequence converging uniformly on compact sets of $\mathbb{R}^8$ to a function f, solution of the minimal surface equation in $\mathbb{R}^8$. The function f satisfies

$$|f(x)| \geq |f_1(x)| \qquad \text{in } \mathbb{R}^8$$

and therefore

$$\limsup_{x \to +\infty} |f(x)| \, |x|^{-3} \geq 1$$

This implies that the graph of f cannot be a hyperplane, so that the Bernstein theorem cannot hold in $\mathbb{R}^8$.

It remains to prove (17.15) and (17.16). We split the minimal surface operator into two parts: $\mathscr{E} = \mathscr{E}_0 + \mathscr{D}$, where

$$\mathscr{E}_0(F) = F_v^2 F_{uu} - 2F_u F_v F_{uv} + F_u^2 F_{vv} + 3\left(\frac{F_u}{u} + \frac{F_v}{v}\right)(F_u^2 + F_v^2)$$

and

$$\mathscr{D}(F) = F_{uu} + F_{vv} + 3\left(\frac{F_u}{u} + \frac{F_v}{v}\right)$$

If we define two new independent variables:

$$r = u^2 + v^2$$

$$t = \frac{u^2 - v^2}{u^2 + v^2}$$

and we write $F(u, v) = G(r, t)$, we have

$$(17.19)\ \mathscr{E}_0(F) = 8(1 - t^2)\left\{2G_t^2 G_{rr} - 4G_t G_r G_{rt} + 2G_r^2 G_{tt} + \right.$$

$$\left. + \frac{9}{r}G_r G_t^2 - \frac{6t}{r^2}G_t^3 + 7rG_r^3 - 8tG_t G_r^2\right\}$$

$$(17.20)\ \mathscr{D}(F) = 64\left(G_r - \frac{t}{r}G_t\right) + 16r\left[G_{rr} + \frac{1}{r^2}(1 - t^2)G_{tt}\right].$$

Taking $G(r, t) = tr^{3/2}$ we obtain:

$$\frac{1}{8}\mathscr{E}_0(F) = \frac{45}{8}r^{5/2}t^3$$

$$\frac{1}{8}\mathscr{D}(F) = \frac{11}{2}r^{1/2}t$$

and hence (17.15) are satisfied since $t > 0$ in $\mathring{T}_1$ and $t < 0$ in $\mathring{T}_2$. The construction of F_2 is more involved. We begin with the choice

$$G(r, t) = rt + r^{3/2}g(t).$$

We have:

$$(17.21) \quad \frac{1}{8}\mathscr{E}_0(F) = r^{5/2}M_1(t) + r^2 M_2(t) + r^{3/2}M_3(t) - rt$$

where

$$(17.22) \quad M_1(t) = (1 - t^2)\left(6gg'^2 - 6tg'^3 + \frac{9}{2}g^2 g''\right) + \frac{189}{8}g^3 - 18tg^2 g'$$

$$(17.23) \quad M_2(t) = (1 - t^2)(15gg' - 15tg'^2 + 6tgg'') + \frac{117}{4}tg^2 - 24t^2 gg'$$

$$(17.24) \quad M_3(t) = (1 - t^2)(9g - 10tg' + 2t^2 g'') - 8t^3 g' + \frac{15}{2}t^2 g.$$

We choose now

$$g(t) = t(1 + A|t|^{\lambda - 1}); \qquad A > 0,$$

and we note that it is sufficient to prove the first inequality (17.16) since G is an odd function of t. We can therefore suppose $0 < t \leq 1$. Writing M_1 as a polynomial in A we get

$$M_1(t) = A^3 t^{3\lambda - 2}\left\{2\lambda(\lambda - 1)\left(\frac{9}{4} - 3\lambda\right)(1 - t^2) + \frac{9}{4}\left(\frac{21}{2} - 8\lambda\right)t^2\right\} +$$

$$+ A^2 t^{2\lambda - 1}\left\{-3\lambda(\lambda - 1)(1 - t^2) + \frac{9}{4}\left[\frac{63}{2} - 8(1 + 2\lambda)\right]t^2\right\} +$$

$$+ At^\lambda\left\{2(\lambda - 1)\left(\frac{9}{4}\lambda - 3\right)(1 - t^2) + \frac{9}{4}\left[\frac{63}{2} - 8(2 + \lambda)\right]t^2\right\} +$$

$$+ \frac{45}{8}t^3.$$

Since $\frac{21}{16} < \lambda < \frac{4}{3}$ we have $\frac{9}{4}\lambda - 3 < 0$, $\frac{21}{2} - 8\lambda < 0$, and of course $\frac{9}{4} - 3\lambda < 0$, so that

$$M_1(t) \leq -c_1(A^3 t^{3\lambda - 2} + A^2 t^{2\lambda - 1} + At^\lambda) + c_2(A^2 t^{2\lambda + 1} + At^{\lambda + 2} + t^3)$$

for some $c_1, c_2 > 0$.

Recalling that $0 < t \leq 1$ and taking A large enough ($A \geq c_3$) we find

$$M_1(t) \leq -c_4(A^3 t^{3\lambda-2} + A^2 t^{2\lambda-1} + At^\lambda).$$

In a similar way:

$$M_2(t) \leq -c_5(A^2 t^{2\lambda-1} + At^\lambda)$$

$$M_3(t) \leq -c_6(At^\lambda + t)$$

and in conclusion:

$$(17.25) \quad \frac{1}{8} \mathscr{E}_0(F) \leq -c_7(r^{5/2} t^\lambda + rt)$$

in $\mathring{T}_1$, provided $A \geq c_3$.

A similar estimate gives

$$(17.26) \quad \frac{1}{8} \mathscr{D}(F) \leq c_8 r^{1/2} t^{\lambda-2}.$$

Now let $H(z)$ be a monotone increasing function in $\mathbb{R}$. Then identically:

$$\mathscr{E}_0(H(F)) = H'(F)^3 \cdot \mathscr{E}_0(F)$$

and

$$\mathscr{D}(H(F)) = H'(F)\mathscr{D}F + 16rH''(F)\left(G_r^2 + \frac{1-t^2}{r^2}G_t^2\right).$$

In our case we have

$$r\left(G_r^2 + \frac{1-t^2}{r^2}G_t^2\right) \geq r + r^2$$

and therefore, if we suppose that

$$(17.27) \quad H'(z) \geq 1; \qquad H''(z) \leq 0$$

we obtain from (17.25) and (17.26):

$$\mathscr{E}(H(F)) \leqq -c_7(r^{5/2}t^\lambda + rt)H'(F)^3 + c_8 r^{1/2}t^{\lambda-2}H'(F) + 16(r+r^2)H''(F).$$

If $r^2 t^2 \geqq c_9 = c_8/c_7$ we have immediately $\mathscr{E}(H(F)) \leqq 0$ from (17.27). We can therefore restrict ourselves to the domain

$$(17.28)\ r^2 t^2 < c_9.$$

In order to show that $\mathscr{E}(H(F)) \leqq 0$ in this domain too, it is sufficient to prove that $\mathscr{D}(H(F)) \leqq 0$ there, and hence that

$$(17.29)\quad -\frac{H''(F)}{H'(F)} \geqq \frac{c_{10}}{(r^{1/2} + r^{3/2})t^{2-\lambda}}$$

When $r \geqq 1$ we have $F \leqq c_{11}r^{3/2}t$ and therefore

$$F^{2-\lambda} + F^{2\lambda-1} \leqq c_{12}[(r^{3/2}t)^{2-\lambda} + (r^{3/2}t)^{2\lambda-1}] \leqq$$

$$\leqq c_{12}[r^{3/2}t^{2-\lambda} + r^{3/2}t^{2-\lambda}(rt)^{3\lambda-3}] \leqq$$

$$\leqq c_{13}r^{3/2}t^{2-\lambda} \leqq c_{13}(r^{1/2} + r^{3/2})t^{2-\lambda}$$

since $r^2 t^2 \leqq c_9$.

The same inequality holds of $r \leqq 1$. Actually in this case we have $F \leqq c_{14}rt \leqq c_{14}$ and therefore

$$F^{2-\lambda} + F^{2\lambda-1} \leqq c_{15}F^{2-\lambda} \leqq c_{16}(rt)^{2-\lambda} \leqq c_{16}(r^{1/2} + r^{3/2})t^{2-\lambda}$$

since $2 - \lambda > 1/2$.
  In conclusion, if

$$(17.30)\quad -\frac{H''(z)}{H'(z)} \geqq \frac{c_{17}}{z^{2-\lambda} + z^{2\lambda-1}}$$

for some large c_{17}, and if $H'(z) \geqq 1$, then (17.29) holds, and therefore $\mathscr{E}(H(F)) < 0$ in $\mathring{T}_1$.
  The inequality (17.30) is satisfied if we take

$$H'(z) = \exp\left(c_{17}\int_z^{+\infty} \frac{dt}{t^{2-\lambda} + t^{2\lambda-1}}\right),$$

an admissible choice since (noting that $2 - \lambda < 1$ and $2\lambda - 1 > 1$)

$$\int_0^{+\infty} \frac{dt}{t^{2-\lambda} + t^{2\lambda-1}} < +\infty$$

and

$$H(z) = \int_0^z H'(w)dw.$$

With this choice, the function

$$F_2 = H(F)$$

satisfies the inequalities (17.26).

We have in conclusion:

17.10   Theorem [B DG G]: *Let $n \geq 8$. Then there exist entire solutions of the minimal surface equation*

$$D_i\left(\frac{D_i u}{\sqrt{1+|Du|^2}}\right) = 0$$

which are not hyperplanes.

Proof: We have constructed such a function f for $n = 8$. If $n > 8$ it is sufficient to take

$$u(x_1, \ldots, x_n) = f(x_1, \ldots, x_8). \qquad \square$$

This result and Theorem 17.8 give a complete solution for the Bernstein problem.

Appendix A

A.1 De La Vallée Poussin Theorem: *Suppose $\{\mu_k\}$ is a sequence of non-negative Radon measures of uniformly bounded total variation, that is*

(A.1) $\mu_k(\mathbb{R}^n) \leq M$ *for all k.*

Then there exists a Radon measure μ and a subsequence $\{v_j\} = \{\mu_{k_j}\}$ such that, for every $g \in C_0^0(\mathbb{R}^n)$

(A.2) $\mu(g) = \lim_{j \to \infty} v_j(g)$.

Moreover, for every bounded Borel set E such that $\mu(\partial E) = 0$, we have

(A.3) $\mu(E) = \lim_{j \to \infty} v_j(E)$.

Proof: It is well known that $C_0^0(\mathbb{R}^n)$ is separable; that is, there exists a countable set $\{g_i\} \subseteq C_0^0(\mathbb{R}^n)$ such that for every $g \in C_0^0(\mathbb{R}^n)$ there exists a subsequence $\{g_{i_k}\}$ with $g_{i_k} \to g$ uniformly. By (A.1) the sequence $\mu_j(g_1)$ is bounded and hence we can extract a subsequence $\{\mu_j^1(g_1)\}$ convergent to a real number a^1. Moreover, we may suppose that

$$|\mu_j^1(g_1) - a^1| < \frac{1}{j}.$$

Now $\mu_j^1(g_2)$ is bounded and we can extract a subsequence $\{\mu_j^2(g_2)\}$ such that

$$|\mu_j^2(g_2) - a^2| < \frac{1}{j}.$$

Proceeding by induction, given $\{\mu_j^k\}$ we can extract a subsequence $\{\mu_j^{k+1}\}$ such that

$$|\mu_j^{k+1}(g_{k+1}) - a^{k+1}| < \frac{1}{j}.$$

If $v_j = \mu_j^j$, then we have for $j > k$

$$|v_j(g_k) - a^k| < \frac{1}{j} \qquad k = 1, 2, \ldots$$

and hence, for every $k \in \mathbb{N}$,

$$\lim_{j \to \infty} v_j(g_k) = a^k.$$

We define the measure μ as

$$\mu(g_k) = a^k.$$

From (A.1)

$$|\mu(g_k)| = \lim_{j \to \infty} |v_j(g_k)| \leq M \sup |g_k|$$

and hence μ can be extended as a continuous linear functional on $C_0^0(\mathbb{R}^n)$ (using the density of $\{g_i\}$).

If $g \in C_0^0(\mathbb{R}^n)$, let $g_{j_k} \to g$ uniformly. Then

$$|v_i(g) - \mu(g)| \leq |v_i(g_{j_k}) - \mu(g_{j_k})| + |v_i(g) - v_i(g_{j_k})| + |\mu(g_{j_k}) - \mu(g)|$$

$$\leq |v_i(g_{j_k}) - \mu(g_{j_k})| + 2M \sup |g - g_{j_k}|$$

and hence

$$\limsup_{i \to \infty} |v_i(g) - \mu(g)| \leq 2M \sup |g - g_{j_k}|.$$

Letting $k \to \infty$, we see that (A.2) follows.

To prove (A.3) we first observe that for any open set A we have

$$\mu(A) = \sup \{\mu(g) : g \in C_0^0(A), |g| \leq 1\}$$

and if C is compact

$$\mu(C) = \inf \{\mu(g) : g \in C_0^0(\mathbb{R}^n), 0 \leq g \leq 1, g = 1 \text{ on } C\}.$$

Let $\overset{\circ}{E}$ = Interior of E, and suppose $g \in C_0^0(\overset{\circ}{E})$ is such that $|g| \leq 1$. Then we have

$$\mu(g) = \lim_{j \to \infty} v_j(g) \leq \liminf_{j \to \infty} v_j(\overset{\circ}{E})$$

and hence

(A.4) $\mu(\overset{\circ}{E}) \leq \liminf_{j \to \infty} v_j(\overset{\circ}{E})$

On the other hand, if $0 \leq g \leq 1$ and $g = 1$ on $\bar{E}$,

$$\mu(g) = \lim_{j \to \infty} v_j(g) \geq \limsup_{j \to \infty} v_j(\bar{E})$$

so that

(A.5) $\mu(\bar{E}) \geq \limsup_{j \to \infty} v_j(\bar{E})$.

Combining (A.4) and (A.5) shows that (A.3) holds.

A.2   Theorem A.1 for vector-valued measures

Suppose now that $\{\vec{\mu}_j\}$ is a sequence of vector-valued Radon measures, so that

$$\vec{\mu}_j : C_0^0(\mathbb{R}^n) \to \mathbb{R}^m.$$

Suppose that the total variations of the $\vec{\mu}_j$ are uniformly bounded, that is

$$|\vec{\mu}_j|(\mathbb{R}^n) \leq M,$$

and that, for every $g \in C_0^0(\mathbb{R}^n)$, the sequence $\{\vec{\mu}_j(g)\}$ converges to a vector denoted $\vec{\mu}(g)$. (This convergence assumption is not restrictive as, given the uniform boundedness we can always extract a subsequence of $\{\vec{\mu}_j\}$ as in A.1 which has the required convergence properties.)

Let μ_j^α, $(\alpha = 1, \ldots, m)$ be the scalar components of $\vec{\mu}_j$ and let $\mu_j^{\alpha+}$, $\mu_j^{\alpha-}$ be the positive and negative part of μ_j^α. We have

$$\mu_j^{\alpha\pm}(\mathbb{R}^n) \leq |\vec{\mu}_j|(\mathbb{R}^n) \leq M$$

and hence, by A.1, for every $\alpha = 1, 2, \ldots, m$ we can find Radon measures $\mu^{\alpha+}$ and $\mu^{\alpha-}$ such that

$$\mu_j^{\alpha+}(g) \to \mu^{\alpha+}(g)$$

$$\mu_j^{\alpha-}(g) \to \mu^{\alpha-}(g) \qquad \text{for every } g \in C_0^0(\mathbb{R}^n).$$

If we define

$$\mu(g) = \mu^+(g) - \mu^-(g),$$

where $\mu^+(g)$ is the vector with components $\mu^{\alpha+}(g)$ and $\mu^-(g)$ is the vector with components $\mu^{\alpha-}(g)$, μ is as a vector-valued Radon measure. Moreover, for every bounded Borel set E such that

$$\mu^{\alpha+}(\partial E) = \mu^{\alpha-}(\partial E) = 0 \qquad \alpha = 1, \ldots, m,$$

we have

$$\mu(E) = \lim_{j \to \infty} \mu_j(E).$$

Thus if we call μ^* the positive measure

$$\mu^*(E) = |\mu^+ + \mu^-|(E)$$

we have

$$\mu(E) = \lim_{j \to \infty} \mu_j(E)$$

for every bounded Borel set E such that $\mu^*(\partial E) = 0$. Clearly $|\mu|(E) \leq \mu^*(E)$.

A.3   Theorem A.1 for Caccioppoli Sets

Now consider A.2 in the particular case where F_j are Caccioppoli sets in $\mathbb{R}^n$ and

$$\mu_j(E) = \int_E D\varphi_{F_j}.$$

Suppose that

$$\lim_{j \to \infty} \varphi_{F_j} = \varphi_F \qquad \text{in } L^1_{\text{loc}}(\mathbb{R}^n)$$

and that

$$\int |D\varphi_{F_j}| \leq M.$$

As in A.1 we may extract a subsequence from $\{F_j\}$ such that $D\varphi_{F_{j_k}}$ converges to some Radon measure μ. Then by the definition of $D\varphi_F$ we must have that

$$\mu(g) = D\varphi_F(g) \qquad \text{for } g \in C^1_0(\mathbb{R}^n).$$

Since $C^1_0(\mathbb{R}^n)$ is dense in $C^0_0(\mathbb{R}^n)$ we have that

$$\mu = D\varphi_F$$

and

$$D\varphi_{F_j}(g) \to D\varphi_F(g) \qquad \text{for } g \in C^0_0(\mathbb{R}^n).$$

We may also conclude, as in A.2, that there exists a Radon measure μ^* such that, for every set E with $\mu^*(\partial E) = 0$,

$$\lim_{j \to \infty} \int_E D\varphi_{F_j} = \int_E D\varphi_F.$$

Moreover we can show

$$|D\varphi_F| \leq \mu^*.$$

In particular, as μ^* is a Radon measure we have, for almost all ρ, $\mu^*(\partial B_\rho) = 0$ and so

$$\lim_{j \to \infty} \int_{B_\rho} D\varphi_{F_j} = \int_{B_\rho} D\varphi_F \qquad \text{for almost all } \rho.$$

Appendix B.   The Distance Function

We shall gather here some results concerning the distance function:

$$d(x) = \text{dist}\,(x,\ \partial\Omega).$$

For a detailed discussion see [GT].

Let Ω be a bounded open set with C^k boundary $\partial\Omega$, $k \geq 2$. It is clear that for every point $y \in \partial\Omega$ there exists a ball $B \subset \Omega$ such that $\bar{B} \cap \partial\Omega = \{y\}$. Since $\partial\Omega$ is compact, the radius of the biggest ball with this property is bounded below by some positive constant R_0 independent of y. It is easily seen that R_0^{-1} gives an upper bound for the principal curvatures of $\partial\Omega$.

If we set for $t > 0$:

$$S_t = \{x \in \Omega : d(x) < t\}$$

$$\Gamma_t = \{x \in \Omega : d(x) = t\}$$

then for $t < R_0$ every point $x \in S_t$ has a unique point at least distance in $\partial\Omega$; i.e. a point $y(x)$ such that $d(x) = |x - y(x)|$. The points x and y are related by

$$x = y + d(x)\ v\ (y)$$

where $v(y)$ is the inner normal unit vector to $\partial\Omega$ at y.

Now let $y_0 \in \partial\Omega$; we can suppose that $y_0 = 0$ and that the tangent plane to $\partial\Omega$ at y_0 is the horizontal plane $x_n = 0$. Then in a neighborhood V of $0 \in \mathbb{R}^{n-1}$, $\partial\Omega$ is the graph of a function $y_n = f(\bar{y})$, $\bar{y} = (y_1, y_2, \ldots, y_{n-1})$, such that $Df(0) = 0$.

With a possible rotation around the x_n-axis we may suppose that the matrix $D^2f(0)$ is diagonal; if we suppose that Ω lies below the graph of f we have

(B.1) $D^2f(0) = \text{diag}\,[-k_1,\ -k_2,\ \ldots,\ -k_{n-1}]$

where $k_1, \ldots, k_{n-1}$ are the principal curvatures of $\partial\Omega$ at 0 (we choose the sign in such a way that the principal curvatures are ≥ 0 if Ω is convex).

For $(\bar{y}, d) \in V \times \mathbb{R}$ we define

$$g(\bar{y}, d) = y + v(y)d, \qquad y = (\bar{y}, f(\bar{y}));$$

we have $g \in C^{k-1}$, and recalling that

(B.2) $v_n(y) = -(1 + Df(\bar{y})^2)^{-1/2}; \quad v_\alpha(y) = -v_n(y)D_\alpha f(\bar{y})(\alpha = 1, 2, \ldots, n-1)$

we have from (B.1):

(B.3) $Dg(0, d) = \text{diag}[1 - k_1 d, 1 - k_2 d, \ldots, 1 - k_{n-1}d].$

If $d < R_0$ the determinant of the Jacobian matrix $Dg(0, d)$ is non-zero and therefore we may write y and d as C^{k-1}-functions of x in a neighborhood of $x_d = (0, d)$, and hence for $x \in S_{R_0}$. Moreover $Dd(x)$ is the unit normal vector to $\Gamma_{d(x)}$ at x, and hence

(B.4) $Dd(x) = v(y(x)) = v(x)$

Since the right-hand side of (B.4) is a C^{k-1}-function, we conclude that $d(x)$ is of class C^k in S_{R_0}. Also from (B.4) it follows that the mean curvature of the surface $\Gamma_{d(x)}$ at x is given by:

$$-\frac{1}{n-1}D_i v_i(x) = \frac{-1}{n-1}\Delta d(x).$$

We can estimate the last quantity using again (B.4). We have:

$$D_i D_j d(x) = D_i[v_j(y(x))] = D_h v_j(y(x)) \cdot D_i y_h(x)$$

and if we suppose that $y(x) = 0$ and that (B.1) holds we find, taking into account (B.2):

$$\Delta d(x) = -\sum_{i=1}^{n-1} k_i/(1 - k_i d(x)).$$

We remark in particular that Δd decreases when x moves along the normal towards the interior of Ω, and therefore if the mean curvature of $\partial\Omega$ is non-negative we have $\Delta d \leq 0$ in S_{R_0}.

Appendix C. Elliptic Equations of the Second Order

We shall state here the principal results, concerning in particular the regularity of the solutions of elliptic partial differential equations of second order, that we have used throughout these notes, giving where it is possible an outline of the proofs. For a more detailed discussion we refer to standard books on the subject; we mention in particular [GT], [LU2] and [GE4].

We consider weak solutions of the equation

(C.1) $\operatorname{div} A(x,\, Du) = 0$ in Ω;

namely, functions $u \in W^{1,2}_{loc}(\Omega)$ such that for every $\varphi \in C^\infty_0(\Omega)$:

(C.2) $\int A_i(x,\, Du) D_i \varphi \, dx = 0.$

The vector field $A(x, p)$ is supposed to satisfy the inequalities:

(C.3) $|A(x, p)| \leqq c_1 |p| + |f(x)|$

(C.4) $\nu |\xi|^2 \leqq \displaystyle\sum_{i,j=1}^{n} \frac{\partial A_i}{\partial p_j} \xi_i \xi_j \leqq \Lambda |\xi|^2 \qquad \forall \xi \in \mathbb{R}^n$

with $f \in L^2_{loc}(\Omega)$ and $\nu > 0$.

In particular we shall be interested in the case

$$A(p) = T(p) = p/(1 + |p|^2)^{1/2}$$

(see Chapter 12), or else

$$A_i(x, p) = a_{ij}(x)p_j - f_i(x)$$

with $a_{ij} = a_{ji}$. In both cases we have

$$\frac{\partial A_i}{\partial p_j} = \frac{\partial A_j}{\partial p_i}$$

an assumption that, although not necessary, will sometimes simplify the proofs. We note that in virtue of (C.3), equation (C.2) holds for every $\varphi \in W^{1,2}(\Omega)$ with compact support.

C.1. Theorem: *Let* $u \in W_{loc}^{1,2}(\Omega)$ *be a weak solution of equation (C.2), and suppose that (C.3) and (C.4) are satisfied, and moreover*

(C.5) $|\partial A_i/\partial x_j(x, p)| \leq c_1 |p| + |f(x)|.$

Then $u \in W_{loc}^{2,2}(\Omega)$, *and every derivative* $w = D_s u$ *satisfies the equation*

(C.6) $\int_{\Omega} \{a_{ij}(x)D_j w + b_i(x)\} D_i \varphi \, dx = 0, \qquad \forall \varphi \in C_0^{\infty}(\Omega)$

where

$$a_{ij}(x) = \frac{\partial A_i}{\partial p_j}(x, Du(x)); \qquad b_i = \frac{\partial A_i}{\partial x_s}(x, Du(x)).$$

Proof: We use the difference quotients:

$$\tau_{h,s} u(x) = h^{-1}[u(x + he_s) - u(x)]$$

where e_s is the unit vector in the x_s-direction. If $\eta \in C_0^{\infty}(\Omega)$ and $|h| < \frac{1}{2} \text{dist}(\partial\Omega, \text{spt}\,\eta)$ we may take $\varphi = \tau_{-h}(\eta^2 \tau_h u)$ in (C.2). If we remark that

$$\int f \tau_{-h} g \, dx = -\int g \tau_h f \, dx$$

we obtain

(C.7) $\int \tau_h A_i(x, Du) D_i(\eta^2 \tau_h u) dx = 0,$

where we have suppressed the index s, and we sum over repeated indices (not over h, of course!). We have:

$$\tau_h A_i(x, Du) = h^{-1} \int_0^1 \frac{d}{dt} A_i(x + the_s, Du + th\tau_h Du) dt = \theta_{ij} D_j \tau_h u + g_i$$

where

$$\theta_{ij} = \int_0^1 \frac{\partial A_i}{\partial p_j}(x + the_s, Du + th\tau_h Du) dt$$

$$g_i = \int_0^1 \frac{\partial A_i}{\partial x_s}(x + the_s, Du + th\tau_h Du) dt.$$

Inserting these in (C.7) we find with simple calculations (here we use the simplifying assumption $\theta_{ij} = \theta_{ji}$):

$$\int \theta_{ij} D_j(\eta v) D_i(\eta v) dx = \int \theta_{ij} v^2 D_i \eta D_j \eta dx - \int \eta g_i D_i(\eta v) dx - \int \eta g_i v D_i \eta dx$$

where we have set $v = \tau_h u$.

Using (C.4) and the inequality $2ab \leqq \varepsilon a^2 + \varepsilon^{-1} b^2$, $\varepsilon > 0$, we easily obtain

$$\int |D(\eta v)|^2 dx \leqq c_2 \{ \int v^2 |D\eta|^2 dx + \int g^2 \eta^2 dx \}.$$

Consider now four open sets $\Omega_1 \subset\subset \Omega_2 \subset\subset \Omega_3 \subset\subset \Omega_4 = \Omega$, and let $|h|$ be smaller than each of the distances $\text{dist}(\Omega_i, \partial\Omega_{i+1})$. Let $\eta \in C_0^\infty(\Omega_2)$ with $\eta = 1$ on Ω_1. We have:

$$\int\limits_{\Omega_1} |Dv|^2 dx \leqq c_3 \{ \int\limits_{\Omega_2} v^2 dx + \int\limits_{\Omega_2} g^2 dx \}.$$

On the other hand, since $u \in W_{loc}^{1,2}(\Omega)$ we have (see e.g. [GE4])

$$\int\limits_{\Omega_2} v^2 dx \leqq c_4 \int\limits_{\Omega_3} |Du|^2 dx;$$

moreover from (C.5) we can easily show:

$$\int\limits_{\Omega_2} g^2 dx \leqq c_5 \int\limits_{\Omega_3} (|Du|^2 + f^2) dx$$

and in conclusion

$$\int\limits_{\Omega_1} |Dv|^2 dx \leqq c_6(\Omega_1).$$

The last estimate implies (see [GE4]) that $u \in W^{2,2}(\Omega_1)$, and therefore, since Ω_1 is arbitrary, $u \in W_{loc}^{2,2}(\Omega)$. Finally, taking $\varphi = D_s\psi$ in (C.2) and integrating by parts we have at once (C.6)   □

The next theorem is one of the cornerstones of the theory of elliptic partial differential equations. For its proof we refer to [MJ] (see also [GE4]).

C.2   **Theorem** *(Harnack's inequality): Let $u \in W_{loc}^{1,2}$ be a positive weak solution of the equation*

(C.8) $D_i(a_{ij}(x)D_j u) = 0$

in a ball B_R, with bounded measurable coefficients a_{ij} satisfying the inequality

(C.9) $a_{ij}\xi_i\xi_j \geq v|\xi|^2, \qquad v > 0.$

Then for every $\alpha < 1$ there exists a constant c depending on α, but not on u, such that:

(C.10) $\sup_{B_{\alpha R}} u \leq c(\alpha) \inf_{B_{\alpha R}} u$

As a consequence of the Harnack's inequality we can prove the regularity theorem of De Giorgi [DG3] and Nash [NJ]:

C.3   **Theorem:** *Let u be a solution of equation (C.8) with conditions (C.9). Then u is Hölder-continuous in Ω.*

Proof: Let $x_0 \in \Omega$ and let $B_R(x_0) \subset\subset \Omega$. For $r < R$ we set:

$$m(r) = \inf_{B_r(x_0)} u$$

$$M(r) = \sup_{B_r(x_0)} u$$

$$\omega(r) = M(r) - m(r)$$

The functions

$$v(x) = u(x) - m(r)$$

$$w(x) = M(r) - u(x)$$

are positive solutions of (C.8) in $B_r(x_0)$. From (C.10) with $\alpha = 1/2$ we have therefore

$$M(r/2) - m(r) \leq c\{m(r/2) - m(r)\}$$

$$M(r) - m(r/2) \leq c\{M(r) - M(r/2)\}.$$

Adding these two inequalities we find:

$$\omega(r) + \omega(r/2) \leq c\{\omega(r) - \omega(r/2)\}$$

and hence

$$\omega(r/2) \leq \frac{c-1}{c+1}\omega(r).$$

If we set $\theta = \dfrac{c-1}{c+1}$ we obtain by induction

$$\omega(2^{-k}R) \leq \theta^k\omega(R)$$

and therefore, since ω is increasing:

$$\omega(r) \leq \theta^{-1}(r/R)^{-\log\theta/\log 2}\omega(R).$$

This inequality implies that $u \in C^{0,\gamma}$, with $\gamma = -\log\theta/\log 2$. $\square$

A second consequence of Harnack's inequality is the strong maximum principle.

C.4 Theorem: *Let u be a weak solution of the elliptic equation (C.8) in a connected open set Ω. If u has an interior minimum point it is constant in Ω.*

Proof: Let $m = \min u = u(x_0)$, and let $B_{2R} \subset \Omega$ be a ball centred at x_0. For $\varepsilon > 0$, $u - m + \varepsilon$ is a positive solution of (C.8) in B_{2R}, and therefore from (C.10) with $\alpha = 1/2$:

$$\sup_{B_R}(u - m + \varepsilon) \leq c\inf_{B_R}(u - m + \varepsilon) \leq c\varepsilon.$$

This implies $m \leq u \leq m + (c-1)\varepsilon$ in B_R and since ε is arbitrary, $u = m$ in B_R. We can therefore conclude that the set

$$\{x \in \Omega : u(x) = m\}$$

is both open and closed, and hence $u(x) = m$ in Ω.   □

The same conclusion holds if u has an interior maximum point, since in this case $-u$ has a minimum at the interior of Ω.

A similar result holds for super and sub-solutions. We shall state it only for supersolutions, i.e. for functions $w \in W^{1,2}_{loc}(\Omega)$ such that

$$(C.11) \quad \int a_{ij}(x) D_j w D_i \varphi dx \geq 0 \qquad \forall \varphi \in C_0^\infty(\Omega), \qquad \varphi \geq 0.$$

We have

C.5   Lemma (*Weak maximum principle*): *Let w be a supersolution in Ω, and let $w \geq 0$ on $\partial\Omega$. Then $w \geq 0$ in Ω.*

Proof: Let

$$A = \{x \in \Omega : w(x) < 0\}.$$

If A is non-empty we may take $\varphi = \max(0, -w)$ in (C.11), obtaining

$$v \int_A |Dw|^2 dx \leq \int_A a_{ij} D_i w D_j w dx \leq 0$$

and hence $Dw = 0$ in A. Since $w = 0$ on ∂A we have $w = 0$ in A, contradicting the definition of A.   □

C.6   Lemma: *Let w be a supersolution in a connected open set Ω, with $w \geq 0$ on $\partial\Omega$.*

If $w(x_0) = 0$ for some $x_0 \in \Omega$, then $w = 0$ in Ω.

Proof: Let $B_R \subset \Omega$ be a ball centred at x_0, and let $u \in W^{1,2}(B_R)$ be the solution of the equation (C.8) taking the values w on ∂B_R. Since $w - u$ is a supersolution we have $w - u \geq 0$ and hence

$$0 \leq u(x) \leq w(x).$$

On the other hand $w(x_0) = 0$ and therefore $u(x_0) = 0$, so that by Theorem C.4 $u = 0$ in B_R. This implies that $w = 0$ on ∂B_R, and since R is arbitrary $w = 0$ in a neighborhood of x_0. The set where $w = 0$ is then open and closed, so that $w = 0$ in Ω.   □

If w is a supersolution and v is a subsolution, $w - v$ is a supersolution and therefore if $w \geq v$ on $\partial\Omega$ and $w(x_0) = v(x_0)$ for some $x_0 \in \Omega$ we conclude that $w = v$ in Ω.

The same result holds for non-linear equations.

As above, a supersolution of (C.1) is a function $w \in W_{loc}^{1,2}(\Omega)$ satisfying

$$\int A_i(x, Dw)D_i\varphi dx \geq 0 \qquad \forall \varphi \in C_0^\infty(\Omega),\ \varphi \geq 0.$$

There is a similar definition for subsolutions.

C.7   Theorem (*Strong maximum principle*): *Let w and v be a super- and a sub-solution of (C.1), with conditions (C.4), and suppose that $w \geq v$ on $\partial\Omega$, and $w(x_0) = v(x_0)$ for some $x_0 \in \Omega$. Then $w = v$.*

Proof: We have

$$\int [A_i(x, Dw) - A_i(x, Dv)]D_i\varphi dx \geq 0 \qquad \forall \varphi \in C_0^\infty(\Omega),\ \varphi \geq 0.$$

On the other hand:

$$A_i(x, Dw) - A_i(x, Dv) = \int_0^1 \frac{d}{dt}A_i(x, Dv + t(Dw - Dv))dt$$

$$= \int_0^1 \frac{\partial A_i}{\partial p_j}(x, Dv + t(Dw - Dv))D_j(w - v)dt = a_{ij}(x)D_j(w - v)$$

with the coefficients

$$a_{ij}(x) = \int_0^1 \frac{\partial A_i}{\partial p_j}(x, Dv(x) + t(Dw(x) - Dv(x)))dt$$

satisfying (C.9). We can therefore apply the preceding lemma to the function $z = w - v$, and conclude that $w = v$ in Ω.   □

The following consequence of Theorem C.2 is also worth stating.

C.8   Theorem (*Liouville's Theorem*): *Let u be an entire solution of equation* (C.8) *in* $\mathbb{R}^n$, *with condition* (C.9). *Suppose that* $u(x) \geq a$ *in* $\mathbb{R}^n$. *Then u is constant.*

The proof of this result is essentially the same as that of Theorem 17.5, and we shall not repeat it here. Of course the same result holds if $u(x) \leq b$ in $\mathbb{R}^n$.

Let us turn now to the non-homogeneous equation

$$(C.12) \quad \int \{a_{ij}(x)D_j u - f_i\} D_i \varphi \, dx = 0 \qquad \forall \varphi \in C_0^\infty(\Omega).$$

We shall investigate the regularity of the solutions in the Sobolev spaces $W^{m,2}$ and in the Hölder space $C^{m,\alpha}$.

C.9   Theorem: *Let* $u \in W_{loc}^{1,2}(\Omega)$ *be a solution of* (C.12), *with the coefficients* a_{ij} *satisfying the ellipticity condition* (C.9). *If the coefficients* a_{ij} *are of class* C^m *and* $f_i \in W_{loc}^{m,2}(\Omega)$, *the solution u belongs to the space* $W_{loc}^{m+1,2}(\Omega)$.

Proof: If $m = 1$ the result follows from Theorem C.1. Suppose it holds for m and let $a_{ij} \in C^{m+1}$, and $f_i \in W_{loc}^{m+1,2}$. The function u is in $W_{loc}^{m+1,2}(\Omega)$, and its derivative $w = D_s u$ satisfies equation (C.6) with

$$b_i = \frac{\partial a_{ij}}{\partial x_s} D_j u - \frac{\partial f_i}{\partial x_s} \in W_{loc}^{m,2}(\Omega).$$

We may therefore conclude that $w \in W_{loc}^{m+1,2}$ and hence $u \in W_{loc}^{m+2,2}(\Omega)$.   □

C.10   Theorem: *Let* $u \in W_{loc}^{1,2}(\Omega)$ *be a solution of* (C.12) *with conditions* (C.9). *Suppose that the coefficients* a_{ij} *and the functions* f_i *are of class* $C^{m,\alpha}(\Omega)$. *Then* $u \in C^{m+1,\alpha}(\Omega)$.

In the case $m = 0$ the result is usually obtained by means of a representation of the solution (see e.g. [LU1], Ch. 3). A different proof, with a method that has proved useful in several circumstances, is due to Campanato [CS], and can be found in [GE4]. The general case follows by induction as above.

In particular, if the coefficients and the f_i's are infinitely differentiable, the solution u will be of class C^∞.

The following regularity result for non-linear elliptic equations is also worth stating. For simplicity we shall only deal with the case where the coefficients A_i are independent of x.

C.11   Theorem: *Let* $u \in W_{loc}^{1,2}(\Omega)$ *be a weak solution of equation* (C.1) *with conditions* (C.2) *and* (C.3), *and with coefficients independent of* x. *Suppose that the functions* A_i *are of class* $C^{m,\alpha}$. *Then* u *is of class* $C^{m+1,\alpha}$.

Proof: By Theorem C.1 the function $w = D_s u$ is a solution of the equation

$$\int a_{ij}(x) D_j w D_i \varphi \, dx = 0 \qquad \forall \varphi \in C_0^\infty(\Omega)$$

with

$$a_{ij}(x) = \frac{\partial A_i}{\partial p_j}(Du(x)).$$

By De Giorgi-Nash theorem C.3, w is Hölder-continuous with some exponent β, and hence $u \in C^{1,\beta}$.

Now let $m = 1$. We have $a_{ij} \in C^{0,\sigma}$ for some $\sigma > 0$, and by Theorem C.10 $u \in C^{2,\sigma}(\Omega)$. But now $a_{ij} \in C^{0,\alpha}$ and, again by Theorem C.10, $u \in C^{2,\alpha}$. The theorem is thus proved if $m = 1$. Suppose now that the result holds for m, and let $A_i(p)$ be of class $C^{m+1,\alpha}$. The inductive step gives $u \in C^{m+1,\alpha}(\Omega)$ and hence the coefficients a_{ij} are in $C^{m,\alpha}(\Omega)$. A further application of Theorem C.10 gives $w \in C^{m+1,\alpha}$, and therefore $u \in C^{m+2,\alpha}(\Omega)$.   □

In particular, if the functions A_i are of class C^∞, the same is true for u. Moreover, if the coefficients are (real) analytic, the solution u will be analytic, a result whose proof can be found in [MCB1].

Similar theorems hold for the boundary regularity of solutions of the Dirichlet problem. If we add to the assumptions of Theorems C.10 and C.11 the hypothesis that $\partial\Omega$ and φ are of class $C^{m+1,\alpha}$, then the solution to the Dirichlet problem with datum φ at the boundary will be of class $C^{m+1,\alpha}(\bar{\Omega})$.

Bibliography

ADAMS, R.A.
    [AR]   Sobolev Spaces. Academic Press, New York, 1975.

ALLARD, W.
    [AW]   On the first variation of a varifold. Ann. of Math. (2) 95 (1972), 417–491.

ALMGREN, F.J., Jr.
    [AF1]   The theory of varifolds – a variational calculus in the large for the k-dimensional area integrand. Mimeographed notes, Princeton, 1965.
    [AF2]   Existence and regularity almost everywhere of solutions to elliptic variational problems among surfaces of varying topological type and singularity structure. Ann. of Math. (2) 87 (1968), 321–391.

ANZELLOTTI, G.
    [AN]   Dirichlet problem and removable singularities for functionals with linear growth. Boll. Un. Mat. Ital. C(5) 18 (1981) 141–159.

ANZELLOTTI, G. – GIAQUINTA, M.
    [AG]   Funzioni BV e tracce. Rend. Sem. Mat. Padova 60 (1978) 1–21.

ANZELLOTTI, G. – GIAQUINTA, M. – MASSARI, U. – MODICA, G. – PEPE, L.
[AGMMP] Note sul problema di Plateau. Mimeographed notes, Pisa, 1974.

BERNSTEIN, S.
    [BS1]   Sur les surfaces définies au moyen de leur courbure moyenne ou totale. Ann. Sci. Ecole Norm. Sup. 27 (1910) 233–256.
    [BS2]   Sur un théorème de géométrie et son application aux équations aux dérivées partielles du type elliptique. Comm. Soc. Math. de Kharkov (2) 15 (1915–1917), 38–45. German transl.: Math. Z. 26 (1927), 551–558.

BOMBIERI, E. – DE GIORGI, E. – GIUSTI, E.
  [BDGG] Minimal cones and the Bernstein problem. Inv. Math. 7 (1969) 243–268.

BOMBIERI, E. – DE GIORGI, E. – MIRANDA, M.
  [BDGM] Una maggiorazione a priori relativa alle ipersuperfici minimali non parametriche. Arch. Rat. Mech. Anal. 32 (1965) 255–267.

CAMPANATO, S.
    [CS]   Equazioni ellittiche del II ordine e spazi $\mathscr{L}^{2,\lambda}$. Ann. Mat. Pura Appl. 69 (1965) 321–382.

COURANT, R. – HILBERT, D.
    [CH]   Methods of Mathematical Physics. I, II. Interscience, New York, 1953, 1962.

DE GIORGI, E.
    [DG1]   Su una teoria generale della misura $(r-1)$-dimensionale in uno spazio ad r dimensioni. Ann. Mat. Pura Appl. (4) 36 (1954), 191–213.
    [DG2]   Nuovi teoremi relativi alle misure $(r-1)$-dimensionali in uno spazio ad r dimensioni. Ricerche Mat. 4 (1955), 95–113.
    [DG3]   Sulla differenziabilità e l'analiticità delle estremali degli integrali multipli regolari. Mem. Acc. Sci. Torino 3 (1957) 25–43.
    [DG4]   Complementi alla teoria della misura $(n-1)$-dimensionale in uno spazio n-dimensionale. Sem. Mat. Scuola Norm. Sup. Pisa, 1960–61. Editrice Tecnico Scientifica, Pisa, 1961.
    [DG5]   Frontiere orientate di misura minima. Sem. Mat. Scuola Norm. Sup. Pisa, 1960–61. Editrice Tecnico Scientifica, Pisa, 1961.
    [DG6]   Una estensione del teorema di Bernstein. Ann. Scuola Norm. Sup. Pisa (3) 19 (1965), 79–85.

DE GIORGI, È. – COLOMBINI, F. – PICCININI, L.C.
  [DGCP] Frontiere orientate di misura minima e questioni collegate. Quad. Scuola Norm. Sup. Pisa No. 1, 1972. Editrice Tecnico Scientifica, Pisa, 1972.

DE GIORGI, E – STAMPACCHIA, G.
    [DGS]   Sulle singolarità eliminabili delle ipersuperficie minimali. Rend. Acc. Lincei 38 (1965) 352–357.

DOUGLAS, J.
[DJ]   Solution of the problem of Plateau. Trans. Amer. Math. Soc. *33* (1931), 263–321.
DUNFORD, N. – SCHWARTZ, J.T.
[DS]   Linear Operators, Part I. Interscience, New York, 1958.
FAVARD, J.
[FJ]   Cours d'analyse de l'École Polytechnique. Tome III: Théorie des équations. Fasc. I.:
         Équations différentielles. Cahiers Scientifiques, Fasc. XXVI. Gauthier-Villars, Paris,
         1962.
FEDERER, H.
[FH1]   Some properties of distributions whose partial derivatives are representable by
         integration. Bull. Amer. Math. Soc. *74* (1968), 183–186.
[FH2]   Geometric Measure Theory. Springer-Verlag, Berlin·Heidelberg·New York, 1969.
[FH3]   The singular set of area minimizing rectifiable currents with codimension one and of
         area minimizing flat chains modulo two with arbitrary codimension. Bull. Amer.
         Math. Soc. *76* (1970), 767–771.
FINN, R.
[F1]   Remarks relevant to minimal surfaces and to surfaces of constant mean curvature.
         J. d'Anal. Math. *14* (1965) 139–160.
FLEMING, W.H.
[FW]   On the oriented Plateau problem. Rend. Circ. Mat. Palermo (2) *11* (1962), 69–90.
FLEMING, W.H. – RISHEL, R.
[FR]   An integral formula for total gradient variation. Arch. Math. *11* (1960), 218–222.
GAGLIARDO, E.
[GA]   Caratterizzazione delle tracce sulla frontiera relative ad alcune classi di funzioni in
         più variabili. Rend. Sem. Mat. Padova *27* (1957) 284–305.
GILBARG, D.
[GD]   Boundary value problems for nonlinear elliptic equations in *n* variables. Symp. on
         Nonlinear Prob. Madison, Wis. April 1962.
GILBARG, D. – TRUDINGER, N.S.
[GT]   Elliptic Partial Differential Equations of Second Order. Springer-Verlag,
         Berlin·Heidelberg·New York, 1977.
GIUSTI, E.
[GE1]   Regolarità parziale delle soluzioni di sistemi ellittici quasi-lineari di ordine arbitrario.
         Ann. Scuola Norm. Sup. Pisa (3) *23* (1969), 115–141.
[GE2]   Boundary behavior of nonparametric minimal surfaces. Ind. Univ. Math. J. *22* (1972)
         435–444.
[GE3]   Minimal surfaces and functions of bounded variation. Notes on Pure Math. *10*,
         Austr. Nat. Univ. Canberra 1977.
[GE4]   Equazioni ellittiche del secondo ordine. Quaderni U.M.I. *6* Pitagora, Bologna 1978.
GIUSTI, E. – MIRANDA, M.
[GM]   Sulla regolarità delle soluzioni deboli di una classe di sistemi ellittici quasi-lineari.
         Arch. Rational Mech. Anal. 31 (1968/69), 173–184.
HAAR, A.
[HA]   Über das Plateausche Problem. Math. Ann. *97* (1927) 124–258.
HALMOS, P.R.
[HP]   Measure Theory. Van Nostrand, Princeton, 1950. Reprint: Springer-Verlag, Berlin ·
         Heidelberg · New York, 1974.
HARDT, R. – SIMON, L.
[HS]   Boundary regularity and embedded solutions for the oriented Plateau problem.
         Ann. of Math. *110* (1979) 439–486.
JENKINS, H. – SERRIN, J.
[JS1]   Variational problems of minimal surface type. II. Boundary value problems for the
         minimal surface equation. Arch. Rat. Mech. Anal. *21* (1966) 321–342.
[JS2]   The Dirichlet problem for the minimal surface equation in higher dimension. J. Reine
         Ang. Math. *229* (1968) 170–187.
LADYŽENSKAJA, O.A. – URAL'CEVA N.N.
[LU1]   Linear and quasilinear elliptic equations. Academic Press, New York 1968.
[LU2]   Local estimates for the gradients of solutions of non-uniformly elliptic and parabolic
         equations. Comm. Pure Appl. Math. *23* (1970) 677–703.

MASSARI, U.
   [MU]   Problema di Dirichlet per l'equazione delle superfici minime con dato infinito. Ann. Univ. Ferrara 23 (1977) 111–141.
MASSARI, U. – MIRANDA, M.
   [MAM]   A remark on minimal cones. Boll. Un. Mat. Ital. (6) 2–A (1983) 123–125.
MIRANDA, M.
   [MM1]   Distribuzioni aventi derivate misure. Insiemi di perimetro localmente finito. Ann. Scuola Norm. Sup. Pisa (3) 18 (1964), 27–56.
   [MM2]   Superfici cartesiane generalizzate ed insiemi di perimetro finito sui prodotti cartesiani. Ann. Sc. Norm. Sup. Pisa 18 (1964) 515–542.
   [MM3]   Sul minimo dell'integrale del gradiente di una funzione. Ann. Scuola Norm. Sup. Pisa (3) 19 (1965), 627–665.
   [MM4]   Un teorema di esistenza e unicirà per il problema dell'area minima in n variabili. Ann. Sc. Norm. Sup. Pisa 19 (1965) 233–249.
   [MM5]   Comportamento delle successioni convergenti di frontiere minimali. Rend. Sem. Mat. Univ. Padova 38 (1967), 238–257.
   [MM6]   Una maggiorazione integrale per le curvature delle ipersuperfici minimali. Rend. Sem. Mat. Univ. Padova 38 (1967), 91–107.
   [MM7]   Un principio di massimo forte per le frontiere minimali e sua applicazione alle superfici di area minima. Rend. Sem. Mat. Padova 45 (1971) 355–366.
   [MM8]   Superfici minime illimitate. Ann. Sc. Norm. Sup. Pisa (4) 4 (1977) 313–322.
   [MM9]   Sulle singolarità eliminabili delle soluzioni dell'equazione delle superfici minime. Ann. Sc. Norm. Sup. Pisa (4) 4 (1977) 129–132.
MORREY, C.B., JR.
   [MCB1] Multiple Integrals in the Calculus of Variations. Springer-Verlag, Berlin · Heidelberg · New York, 1966.
   [MCB2] Partial regularity results for non-linear elliptic systems. J. Math. Mech. 17 (1967/68), 649–670.
MOSER, J.
   [MJ]   On Harnack's theorem for elliptic differential equations. Comm. Pure. Appl. Math. 14 (1961) 577–591.
MUNROE, M.E.
   [MME]   Measure and Integration. Addison-Wesley, Reading, Mass., 1971.
NASH, J.
   [NJ]   Continuity of solutions of parabolic and elliptic differential equations. Amer. J. of Math. 80 (1958) 931–953.
NITSCHE, J.C.C.
   [NI1]   Elementary proof of Bernstein's theorem on minimal surfaces. Ann. of Math. 66 (1957) 543–544.
   [NI2]   On new results in the theory of minimal surfaces. Bull. Amer. Mat. Soc. 71 (1965) 195–270.
RADÒ, T.
   [RT1]   The problem of the least area and the problem of Plateau. Math. Z. 32 (1930), 763–796.
   [RT2]   On the Problem of Plateau. Springer-Verlag, Berlin, 1933.
ROGERS, C.A.
   [RC]   Hausdorff Measures. Cambridge University Press, Cambridge, 1970.
SANTI, E.
   [SE]   Sul problema al contorno per l'equazione delle superfici di area minima su domini limitati qualunque. Ann. Univ. Ferrara 17 (1972) 13–26.
SIMON, L.
   [SL1]   Boundary regularity for solutions of the non-parametric least area problem. Ann. of Math. 103 (1976) 429–455.
   [SL2]   On a theorem of De Giorgi and Stampacchia. Math. Z. 115 (1977) 199–204.
   [SL3]   Boundary behaviour of solutions of the non-parametric least area problem. Bull. Austr. Math. Soc. 26 (1982) 17–27.
SIMONS, J.
   [SJ]   Minimal varieties in riemannian manifolds. Ann. of Math. (2) 88 (1968), 62–105.

SPRUCK, J.
  [SJ]   Infinite boundary value problems for surfaces of constant mean curvature. Arch.
           Rat. Mech. Anal. *49* (1972) 1–31.
STAMPACCHIA, G.
  [SG]   On some regular multiple integral problem in the calculus of variations. Comm. Pure
           Appl. Math. *16* (1963) 383–421.
TRUDINGER, N.S.
  [TN1]   A new proof of the interior gradient bound for the minimal surface equation in n
           dimensions. Proc. Nat. Acad. Sci. USA *69* (1972) 821–823.
  [TN2]   Gradient estimates and mean curvature. Math. Z. *131* (1973) 165–175.
WHITNEY, H.
  [WH]   Analytic extensions of differentiable functions defined in closed sets. Trans. Amer.
           Math. Soc. *36* (1934), 63–89.
WILLIAMS, G.
  [WG]   Boundary regularity of solutions of the non-parametric least area problem. (Preprint)

Subject Index